Optical Activity

Optical Activity: Basic Properties provides a clear explanation of the phenomenon of optical rotation. It covers the macroscopic as well as the microscopic. The connection between the macroscopic and microscopic description is discussed and particularly the relation between various physical parameters is explored. The focus is on the order-of-magnitude of these values and in particular why optical activity is not observed in many dielectric materials. The possible role of quantum-mechanical effects for explaining optical activity is also explored. Optical Activity is written at an intermediate level and only requires knowledge at the junior or senior undergraduate level. Bridging the gap between existing textbooks and the in-depth treatment found in review articles, this book presents as a further-reader resource for physics, chemistry, and engineering students.

Key Features:

- Detailed explanations of how the mathematical formalism relates to underlying physical mechanisms.
- Careful discussion of why optical activity can only be found in systems [molecules] without inversion symmetry.
- Exploration of the underlying quantum-mechanical origin of optical activity.

Ulrich Zürcher is a Physics Professor at Cleveland State University.

Series in Optics and Optoelectronics

Fourier Optics in Image Processing
Neil Collings

Holography
Principles and Applications
Raymond K. Kostuk

An Introduction to Quantum Optics, Second Edition
Photon and Biphoton Physics
Yanhua Shih

Polarized Light and the Mueller Matrix Approach, Second Edition
José J. Gil, Razvigor Ossikovski

Introduction to Holography, Second Edition
Vincent Toal

Optical Coherence Tomography in Dentistry
Scientific Developments to Clinical Applications
Anderson S. L. Gomes, Denise M. Zezell, Cláudia C. B. O. Mota, John M. Girkin

Holography: Principles and Applications, Second Edition
Raymond K. Kostuk

CAD-Based Optical Design with Quadoa
Rafael G. González-Acuña

Anisotropy of Metamaterials: Beyond Conventional Paradigms
Vladimir Mityushev, Tatjana Gric, Radoslaw Kycia, Natalia Rylko

Optical Science and Engineering
Elias Glytsis

Advanced Optical Techniques in Biosciences
Nirmal Mazumder, Yury Kistenev, Igor Lednev

Fundamental Optical Models
William A. Challener

Optical Activity: Basic Properties
Ulrich Zürcher

For more information about this series, please visit: *https://www.crcpress.com/Series-in-Optics-and-Optoelectronics/book-series/TFOPTICSOPT*

Optical Activity

Basic Properties

Ulrich Zürcher

CRC Press
Taylor & Francis Group
Boca Raton London New York

CRC Press is an imprint of the
Taylor & Francis Group, an **informa** business

Designed cover image: CRC Press

First edition published 2027
by CRC Press
2385 NW Executive Center Drive, Suite 320, Boca Raton FL 33431

and by CRC Press
4 Park Square, Milton Park, Abingdon, Oxon, OX14 4RN

CRC Press is an imprint of Taylor & Francis Group, LLC

ISBN: 978-1-032-91016-1 (hbk)
ISBN: 978-1-032-91017-8 (pbk)
ISBN: 978-1-003-56094-4 (ebk)

DOI: 10.1201/9781003560944

Typeset in CMR10 font
by KnowledgeWorks Global Ltd.

Publisher's note: This book has been prepared from camera-ready copy provided by the authors.

To my friends who all contributed in some fashion, and Colby and Charlie for their (very) early morning walks.

Contents

Foreword

Writing a book is challenging. One reason is the length of the project from the idea to the eventual submission. Any long-term commitment comes with a change in 'emotions.' Probably the usual patterns begin highly enthusiastic, with a 'grind' in the middle, and exhaustion toward the end. But the original enthusiasm is (largely) gone. This is different in this case, and the author is even more intrigued about the phenomena of optical activity (also referred to as optical rotatory power) than in the beginning, which is perhaps the most fortunate outcome as an author.

I hope that my readers will share my view and find that optical activity is a truly fascinating topic deserving a closer look.

Ulrich Zürcher
Cleveland, Ohio

Preface

The phenomena of optical activity is easily described. When linearly polarized light travels through certain dielectric material, e.g., corn syrup, the polarization axis (slightly) rotates; the degree of rotation is proportional to the thickness of the sample.

At a macroscopic level, optical properties of dielectric material is described by the index of refraction n such that the speed of light in matter is $v = c/n$ (where $c = 3.00 \times 10^8$ m/s is the speed of light in vacuum [or air]). For light with a fixed wavelength λ (or wave vector $k = 2\pi/\lambda$) the frequency can be written as $\omega = kc/n$. A frequency dependence $n = n(\omega)$ explains the splitting of white light into different colors by a prism; this is referred to as frequency-dispersion, or simply as 'dispersion.' Optical activity corresponds to different index of refraction $n_r \neq n_l$ so that $\omega_r \neq \omega_l$ and the frequency of the polarization axis follows $\Omega = \omega_r - \omega_l$.

It is quite 'obvious' that optical activity has 'something to do' with a certain handedness of the dielectric material. In a phenomenological description, it is, however, a *description* (a spoken or written representation or account of an object – taken from the *Oxford Dictionary*) but not an *explanation* (a reason or justification given for an action or belief – taken from the *Oxford Dictionary*). It turns out that a molecular description of frequency dispersion cannot simply be tweaked in some fashion to explain handedness and thus optical activity.

The seed of this book was planted when an inquiring student majoring in sciences (but not physics) asked the author to explain the origin of optical activity (optical rotatory power). I was unable to answer on the spot, but promised to have the answer the next day. I did expect to find the answer in either Jackson's or Zangwill's *Electrodynamics* texts; much to my surprise, it is not covered in these books (and may explain my apparent lack of knowledge!).

The author was successful to find brief discussions in Feynman's *Lectures on Physics–Volume I* and Hecht's *Optics*. However, the treatment left him more confused as the explanations alluded to several mechanism including a quantum mechanical origin, chirality of molecules, and valence electrons moving along spirals. In situations like this, the author consults Landau-Lifshitz, where its optical activity can indeed be found in Volume VIII. Unfortunately, it did not lead to more clarity; in fact, just the opposite, as Landau introduces the concept of spatial diffusion. These three standard books are written by highly respected authors, which suggests that the different explanations must 'somehow' be related to each other. However, the relationships between the different approaches are far from 'obvious.'

The author ended up telling the student that the explanation of optical activity is rather 'convoluted' and requires concepts from mathematics and physics outside the topics usually covered in an introductory physics sequence. This is an unacceptable situation in my view; this book makes an attempt to remedy this situation. My initial intent was to write a relatively short paper (4-5 pages). While this is a distinct possibility, it would require prior knowledge from physics and mathematics that are covered only in advanced undergraduate and graduate courses. This book is aimed for those readers who have not read *Jackson's Electrodynamics* and texts of that caliber.

Phrases such as 'it can be shown' generally discourage further reading; the guiding principle was to make book reasonably self-contained. That is, I only assume prior knowledge at the introductory level (specifically, general and organic chemistry, introductory physics [calculus or algebra based], and first-semester calculus). This means that some basic material must be included despite the short length of the text.

Chapter 1 is the longest chapter and contains brief outlines of relevant background material. A reader may choose to skip the first chapter in its entirety and come back to it later on as need. Properties of dielectric matter (liquids and gases) is covered in Chapter 2. The discussion shows that optical activity cannot be treated as a 'small' modification of frequency dispersion as dependence of the index of refraction on the handedness of incident light may suggest. In Chapter 3, we discuss the rotation of the electric field due to an external magnetic field, the so-called magneto-optic, or Faraday effect. While this is an *external* mechanism, it provides important clues for an *internal* mechanism relevant for optical activity. The macroscopic description of rotatory power, so-called *spatial* diffusion, is outlined in Chapter 4. It shows that molecules can no longer be treated as 'points' and that their spatial 'extent' (or area) play an important role in optical activity. In Chapter 4, we also discuss that an explanation based on a quantum-mechanical description is necessary; an explanation based on classical mechanics cannot explain the rotation of the electric field. As such, spatial diffusion is a *phenomenological* description of optical activity. We discuss an outline of a quantum-mechanical treatment and a classical model in Chapter 5. Finally, Chapter 6 is an attempt of a possible short answer to my student's question.

Adjectives such as comprehensive or authoritative do not characterize the content of this book, or even its intent. The aim is to provide a road map to the reader and outline how different descriptions are related to each other. In this 'spirit,' no attempt was made to provide an up-to-date literature review. In my view, the reviews by the 'founding fathers' of physical chemistry (Eyring, Kauzmann, Kirkwood, and other) still provide unparalleled insights into optical activity despite them being about 80 years old; as such, we use these reviews as the basis for this book.

1

Preliminaries

In this chapter, we briefly review basics of classical mechanics (Newtonian physics) and electricity and magnetism, as it applies to properties of atoms and molecules. Because some of these topics require mathematical tools not usually covered in courses outside a physics core curriculum, we also sketch the necessary mathematical background.

We review units of physical quantities in both the MKS (meter-kilogram-second) and cgs (centimeter-gram-second) systems. While the former is now the generally accepted unit, much of the older, and very much foundational, literature on optical activity has been written in the first part of the 20th century, when the cgs system was generally preferred. Considerable 'confusion' about units is particularly prevalent in electricity and magnetism.

For the purpose of this book, detailed properties of atoms and molecules are not important. We choose the simplest systems to find *order-of-magnitude* estimate for atomic and molecular quantities: we choose the hydrogen atom and the water (H_2O) and carbon dioxide (CO_2) molecules to illustrate the basic physical principles underlying dielectric properties of dielectric matter; we are particularly interested how macroscopic properties are explained by microscopic properties.

1.1 Mathematical Background

Vectors: A vector $\vec{A}$ is defined by a magnitude $|\vec{A}|$ and direction, e.g., polar angle ('latitude') θ and azimuthal angle ('longitude') ϕ. A vector in three spatial dimensions can be written in component form

$$\vec{A} = A_x\,\hat{x} + A_y\,\hat{y} + A_z\,\hat{z}, \tag{1.1}$$

DOI: 10.1201/9781003560944-1

where $\hat{x}$, $\hat{y}$, and $\hat{z}$ are unit vectors in three orthogonal directions. (Alternatively, we use coordinates ($x = x_1, y = x_2, z = x_3$) and unit vectors $\hat{x} = \hat{e}_1$, $\hat{y} = \hat{e}_2$, and $\hat{z} = \hat{e}_3$). We define two products: (i) the scalar product $C = \vec{A} \cdot \vec{B} = |\vec{A}|\,|\vec{B}| \cos \chi$, where χ is that the angle between two vectors; in component form

$$C = \vec{A} \cdot \vec{B} = A_x B_x + A_y B_y + A_z B_z. \tag{1.2}$$

(alternatively, $C = \sum_{lm} \delta_{lm} A_l B_m$ with the Kronecker delta $\delta_{lm} = 1$ for $l = m$ and $\delta_{lm} = 0$ for $l \neq 0$). Alternatively, $\vec{A} \cdot \vec{B}$ can be written as a matrix product of the row vector $\vec{A}^T$ and the column vector $\vec{B}$. (ii) The vector product $\vec{D} = \vec{A} \times \vec{B}$, where $|\vec{D}| = |\vec{A}|\,|\vec{B}| \sin \chi$ and the direction of $\vec{D}$ is defined via the right-hand rule; in component form

$$D_x = (\vec{A} \times \vec{B})_x = A_y B_z - A_z B_y, \tag{1.3}$$

and similar for D_y and D_z using $x \to y$, $y \to z$, and $z \to x$. (or alternatively, $D_l = (\vec{A} \times \vec{B})_l = \sum_{mn} \epsilon_{\mathrm{LC},lmn} A_m B_n$, where we used the Levi-Civita symbol $\epsilon_{\mathrm{LC},lmn}$ for $\epsilon_{\mathrm{LC},lmn} = 1$ and $\epsilon_{\mathrm{LC},lmn} = -1$ for even and odd permutations of (1,2,3), respectively, and $\epsilon_{\mathrm{LC},lmn} = 0$ if two indices are repeated.[1]

The vector triple product is the vector product of three vectors $\vec{A}$, $\vec{B}$, and $\vec{C}$ can be written as a superposition

$$\vec{A} \times (\vec{B} \times \vec{C}) = (\vec{A} \cdot \vec{C})\vec{B} - (\vec{A} \cdot \vec{B})\vec{C}. \tag{1.4}$$

We consider the coordinate transformation $\vec{r} \to \vec{r'} = -\vec{r}$ and distinguish between *polar* and *axial* vectors: for polar vector (or just 'vector')

$$\vec{V} \to \vec{V'} = -\vec{V}, \tag{1.5}$$

and axial vector

$$\vec{A} \to \vec{A'} = \vec{A}. \tag{1.6}$$

In mechanics, velocity $\vec{v} = d\vec{r}/dt$, acceleration $\vec{a} = d^2\vec{r}/dt^2$, force $\vec{F}$, momentum are polar vectors, whereas torque $\vec{\tau} = \vec{r} \times \vec{F}$, angular momentum $\vec{L} = \vec{r} \times \vec{p}$, and angular velocity $\vec{\omega}$ are axial vectors. We illustrate the character of vectors by considering the torque under inversion, cf. Fig. 1.1. The position and force vectors are axial vectors and change directions $\vec{r} \to -\vec{r}$ and $\vec{F} \to -\vec{F}$ whereas the torque remains unchanged $\vec{\tau} \to \tau$.

---

[1] Usually, the symbol ϵ_{lm} is used for the Levi-Civita symbol; we use a modified symbol as we reserve the symbol ϵ for the electric permeability.

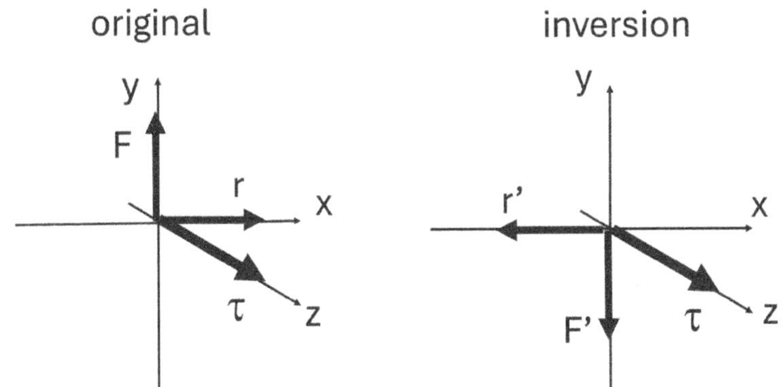

FIGURE 1.1
In the original version, the coordinate vector $\vec{r}$ points along the $+x$-axis, the force $\vec{F}$ points along the $+y$-axis. In the inverted version, the coordinate vector $\vec{r'}$ points along the $-x$-axis, the force $\vec{F}'$ points along the $-y$-axis. The corresponding torques are identical $\vec{\tau} = \vec{r} \times \vec{F} = \vec{r'} \times \vec{F}'$ pointing along the $+z$-axis.

Complex numbers: Complex numbers are defined $z = x + iy$ where the imaginary unit is defined $i = \sqrt{-1}$. The complex conjugate is defined $z^* = x - iy$. The magnitude $|z|^2 = x^2 + y^2 = z \cdot z^*$. The phase is defined $\tan\phi = y/x$, cf. Fig. 1.2. We arrive at the Euler notation

$$z = x + iy = |z|e^{i\phi}. \tag{1.7}$$

In the context of physics, complex notation is a matter of convenience: it is understood that the real part represents the 'physically relevant' solution. We have $de^{i\phi}/d\phi = ie^{i\phi}$ which is equivalent to two equations $d\cos\phi/d\phi = -\sin\phi$ and $d\sin\phi/d\phi = \cos\phi$,

One complex number can be transformed (or mapped) into another complex number by a complex function $z = x + iy \rightarrow f(z) = u(x, y) + iv(x, y)$. The function $f(z)$ is said to be holomorphic when the real and imaginary parts obey the Cauchy-Riemann equations $\partial u/\partial x = \partial v/\partial y$ and $\partial u/\partial y = -\partial v/\partial x$. The (real) integration along the x can be extended to the complex plane along a closed contour γ: $\int_a^b f(x)dx \rightarrow \oint_\gamma f(z)dz$. The Cauchy theorem states that the contour integral vanishes for a holomorphic function $\oint_\gamma f(z)dz = 0$. If $f(z)$ is holomorphic, then $g(z) = f(z)/(z - z_0)$ has

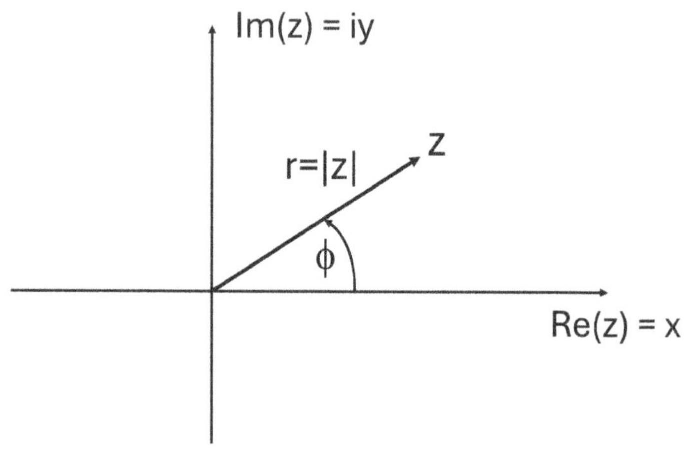

FIGURE 1.2
The complex number $z = x + iy$ in the complex plane with real (imaginary) parts along the horizontal (vertical) axis. The magnitude $|z|$ corresponds to the radius r; the angle ϕ is taken from the real axis $\mathrm{Re}(z)$ in counter-clockwise direction.

a *first-order pole* at $z = z_0$. If z_0 is inside the contour γ, Cauchy's integral theorem states

$$f(z_0) = \frac{1}{2\pi i} \oint \frac{f(z)}{z - z_0} dz, \tag{1.8}$$

where the contour is taken in counter-clockwise orientation. This theorem states that properties of holomorphic functions are characterized by the behavior in the vicinities of poles.

Matrices and tensors: In mechanics, rotation about a fixed axis (chosen as the x-coordinate) is examined; the angular momentum and angular velocity have x-components that are parallel $L_x = I\omega_x$, where I is the moment of inertia. If the axis is not fixed, the angular momentum and angular velocity are no longer parallel. This can easily be observed in the 'wobbly' motion of a spinning egg placed on a table, or the irregular motion of a ball in American football (unless thrown in a 'perfect spiral'). The relation between angular momentum and angular velocity involves all three spatial coordinates and is written as

$$L_x = I_{xx}\omega_x + I_{xy}\omega_y + I_{xz}\omega_z, \tag{1.9}$$

where $(I_{xx}, I_{xy}, ..., I_{yz}, ..., I_{zz})$ are the elements of the *inertia tensor* (or alternatively, $L_l = \sum_m I_{lm}\omega_m$). The 3×3-tensor has nine elements that can be arranged in matrix form. The elements I_{lm} depend on the coordinates and must transform accordingly rotation and inversion.

In linear algebra, Eq. (1.9) is generalized to arbitrary vectors: a vector $\vec{x}$ is transformed into a $\vec{y}$ by the action of an operator $\mathbf{A}$: $\vec{y} = \mathbf{A}\vec{x}$. The vectors $\vec{x}$ and $\vec{y}$ can have arbitrary dimensions N and M. The relation is often written in component form

$$y_m = \sum_{n=1}^{N} A_{mn}x_n, \quad m = 1, 2, \ldots M. \tag{1.10}$$

The two-dimensional array A_{ln} is referred to as a matrix; in the case that the dimensions of the two vectors are the same, the A_{ln} is a square-matrix. The transpose of a square matrix $A^\dagger$ is defined by $(A^\dagger)_{ln} = A_{nl}$. The matrix elements can be complex-valued (i.e., have both real and imaginary parts). A square matrix is Hermitian if $A_{ln} = A^*_{nl}$. An operator $\mathbf{A}$ or matrix A_{mn} defines eigenvectors $\vec{e}$ and eigenvalues λ:

$$\mathbf{A}\vec{e} = \lambda\vec{e}. \tag{1.11}$$

This can be written as $[\mathbf{A} - \lambda\mathbf{I}]\vec{e} = 0$, where $\mathbf{I}$ is the unit operator (or unit matrix $I_{mn} = \delta_{mn}$]: it follows that the determinant must be zero

$$\det(\mathbf{A} - \lambda\mathbf{I}) = 0. \tag{1.12}$$

In matrix notation, this is a polynomial equation of order N (so-called characteristic equation). For a Hermitian matrix (of operator) $\mathbf{A}$, the eigenvalues are real $\lambda = \lambda^*$ and the eigenvector are orthogonal $\vec{e}^{\,*}_m \cdot \vec{e}_n = \delta_{mn}$.

Vector calculus: Roughly speaking, vector calculus is the extension of calculus (i.e., taking derivates) of functions to vector quantities. Here we only discuss vector calculus in cartesian coordinates. For a scalar function (e.g., the potential energy $U = U(x, y, z)$), the gradient is defined as

$$\nabla U = \begin{pmatrix} \partial U/\partial x \\ \partial U/\partial y \\ \partial U/\partial z \end{pmatrix}; \tag{1.13}$$

(or alternatively, $(\nabla U)_i = \partial U/\partial x_i$) this is equivalent with the *steepest descent.* In particular, ∇U is orthogonal (*locally*) to the line of constant $U(x, y, z) = \text{const}$) ('equipotential lines'). For conservative forces (such as the Coulomb force in electrodynamics), we have $\vec{F} = -\nabla U$. The notation $\operatorname{grad} U = \nabla U$ is also used, in particular in older texts. The 'nabla' operator ∇ is a (polar) vector. For a vector field $\vec{A}$, the 'flux' through a surface is defined. The differential version is written in terms of the divergence, which is defined as the scalar product of the nabla operator with a vector function $\vec{A} = A_x(x, y, z)\hat{x} + A_y(x, y, z)\hat{y} + A_z(x, y, z)\hat{z}$:

$$\nabla \cdot \vec{A} = \frac{\partial A_x}{\partial x} + \frac{\partial A_y}{\partial y} + \frac{\partial A_z}{\partial z}, \tag{1.14}$$

where we suppressed the dependence of the components of $\vec{A}$ on the coordinates x, y, z (or alternatively, $\nabla \cdot \vec{A} = \sum_i \partial A_i/\partial x_i$). This is also written as $\operatorname{div}\vec{A}$. The line integral of a vector field around the loop of a finite surface is defined $\oint \vec{A} \cdot d\vec{s}$. The differential version relates to the curl, defined as the vector product of the nabla operator with a vector field $\operatorname{curl}\vec{A}$

$$\nabla \times \vec{A} = \begin{pmatrix} \partial A_z/\partial y - \partial A_y/\partial z \\ \partial A_x/\partial z - \partial A_z/\partial x \\ \partial A_y/\partial x - \partial A_x/\partial y \end{pmatrix}; \tag{1.15}$$

(or alternatively, $(\nabla \times \vec{A})_l = \sum_{mn} \epsilon_{lmn} \partial A_m/x_n$). This is also written as $\operatorname{curl}\vec{A}$. The divergence of gradient is a scalar operator ('Laplacian') $\nabla \cdot \nabla U = \nabla^2 U$:

$$\nabla^2 U = \frac{\partial^2 U}{\partial x^2} + \frac{\partial^2 U}{\partial y^2} + \frac{\partial^2 U}{\partial z^2}. \tag{1.16}$$

(or alternatively, $\nabla^2 U = \sum_i \partial^2 U/\partial x_i^2$). We note that Laplacian of a vector field is defined in component form: $(\nabla^2 \vec{A})_x = \nabla^2 A_x$, and analogous for the y and z components. The curl of a gradient is identically zero $\nabla \times (\nabla U) = 0$; we say that gradient fields are 'curl free.' For the triple product, we have

$$\nabla \times (\nabla \times \vec{A}) = \nabla(\nabla \cdot \vec{A}) - (\nabla \times \nabla)\vec{A}. \tag{1.17}$$

Fourier series and integrals: We consider the interval $-L/2 < x < L/2$ and have a set of orthonormal set of exponentials

$\exp(ik_n x)$, where the wave vectors are $\mathsf{k}_n = 2\pi n/L$. We have the expansion

$$f(x) = \frac{1}{\sqrt{L}} \sum_{n=-\infty}^{\infty} A_n e^{i\mathsf{k}_n x}, \tag{1.18}$$

where the coefficients are given by

$$A_n = \frac{1}{\sqrt{L}} \int_{-L/2}^{L/2} e^{i\mathsf{k}_n x} f(x) dx. \tag{1.19}$$

In particular, for $f(x) = (1/\sqrt{L}) \exp(-i\mathsf{k}_m x)$, we find the Kronecker-delta $A_n = \delta_{nm}$. If the function is even $f(x) = f(-x)$, the function can be represented by a Fourier cosine series: $f(x) = a_0 + 2\sum_{n=1}^{\infty} A_n \cos(\mathsf{k}_n x)$ and if the function is odd $f(x) = -f(-x)$, the function can be represented by a Fourier sine series: $f(x) = 2\sum_{n=1}^{\infty} A_n \sin(\mathsf{k}_n x)$. The wave vector k_n relates to the wavelength $\mathsf{k}_n = 2\pi/\lambda_n$ with $\lambda_n = L/n$. The central feature of Fourier transforms can be seen from Fig. 1.3; finer details of the functions relate to properties at smaller wavelength or larger wave vectors. Put differently, the average behavior is determined by properties at longer wavelengths or smaller wave vectors.

In the limit $L \to \infty$, the discrete wave vector becomes a continuous variable, $\mathsf{k}_n \to \mathsf{k}$, and the Fourier series becomes a Fourier integral

$$f(x) = \frac{1}{\sqrt{2\pi}} \int_{-\infty}^{+\infty} e^{i\mathsf{k}x} \tilde{f}(\mathsf{k}) d\mathsf{k}, \tag{1.20}$$

where

$$\widehat{f}(\mathsf{k}) = \frac{1}{\sqrt{2\pi}} \int_{-\infty}^{+\infty} e^{-i\mathsf{k}x} f(x) dx. \tag{1.21}$$

The orthogonality condition reads $(1/2\pi) \int_{-\infty}^{\infty} e^{i(\mathsf{k}-\mathsf{k}')x} dx = \delta(\mathsf{k} - \mathsf{k}')$.

The Fourier transform can also be used to describe time-dependence: the wave vector k is replaced by the (angular) frequency:

$$f(t) = \frac{1}{\sqrt{2\pi}} \int_{-\infty}^{+\infty} e^{-i\omega t} \tilde{f}(\omega) d\omega, \tag{1.22}$$

where

$$\widehat{f}(\omega) = \frac{1}{\sqrt{2\pi}} \int_{-\infty}^{+\infty} e^{i\omega t} f(t) dt. \tag{1.23}$$

In this context, the coordinate x (time t) and the wave vector k (frequency ω) are referred to a *direct space* and *reciprocal space.* All properties of the function in the direct space $f(x)$ ($f(t)$) have a one-to-one correspondence to properties of the function in the reciprocal space $\widehat{f}(\mathsf{k})$ ($\widehat{f}(\omega)$). The choice of the sign of the phase [$\exp(i\mathsf{k}x)$ and $\exp(-i\omega t)$] is usually used in physics since the overall phase $\exp(i[\mathsf{k}x-\omega t])$ describes a plane wave propagating along the $+x$-direction with speed $c=\omega/\mathsf{k}$.

The Kronecker delta δ_{nm} for integers n and m corresponds to the (Dirac) delta function $\delta(x-x')$ for continuous variables x and x' and is defined as

$$f(x)=\int_{-\infty}^{\infty} f(x')\delta(x-x')dx'. \tag{1.24}$$

The Fourier transform of the delta function follows

$$\widehat{\delta}(\mathsf{k})=\frac{1}{\sqrt{2\pi}}\int_{-\infty}^{\infty}\delta(x-x')e^{i\mathsf{k}x}dx=\frac{1}{\sqrt{2\pi}}e^{i\mathsf{k}x'} \tag{1.25}$$

so that the magnitude is independent of the wave vector $|\widehat{\delta}(\mathsf{k})|=1$.

We consider the derivate $f'(x)=df/dx$. The Fourier transform follows $\widehat{f'}(q) = (1/\sqrt{2\pi})\int_{-\infty}^{\infty}(df/dx)e^{ix}dx = -\int_{-\infty}^{\infty}(de^{i\mathsf{k}x}/dx)$ $f(x)dx=-i\mathsf{k}\int_{-\infty}^{\infty}e^{i\mathsf{k}x}f(x)dx=-i\mathsf{k}\widehat{f}(\mathsf{k})$. Thus, taking the derivative in direct space corresponds to a multiplication in reciprocal space $d/dx \longleftrightarrow i\mathsf{k}$ ($d/dt \longleftrightarrow i\omega$) and analogously for higher orders $d^2/dx^2 \longleftrightarrow -\mathsf{k}^2$ ($d^2/dt^2 \longleftrightarrow -\omega^2$). If direct and reciprocal spaces are interchanged with each other, we get $x \longleftrightarrow id/d\mathsf{k}$ ($t \longleftrightarrow id/d\omega$) and $x^2 \longleftrightarrow -d^2/d\mathsf{k}^2$ ($t^2 \longleftrightarrow -d^2/d\omega^2$).

Convolution (Faltung): In the 'language of time-frequency', the convolution of two functions is defined as

$$h(t)=\langle\chi * F_{\text{ext}}\rangle(t)=\int_0^t \chi(t-s)F_{\text{ext}}(s)ds; \tag{1.26}$$

that is, the response of a system $h(t)$ to a forcing $F(s)$ at previous times $0<s<t$; the function $\chi(t)$ is referred to as a linear response, or Green's function. The Fourier transform reads

$$\widehat{h}(\omega)=\widehat{\chi}(\omega)\,\widehat{F}_{\text{ext}}(\omega), \tag{1.27}$$

where $\widehat{\chi}(\omega)$ is the dynamic susceptibility. We now discuss the frequency dependence of the dynamic susceptibility and, in particular,

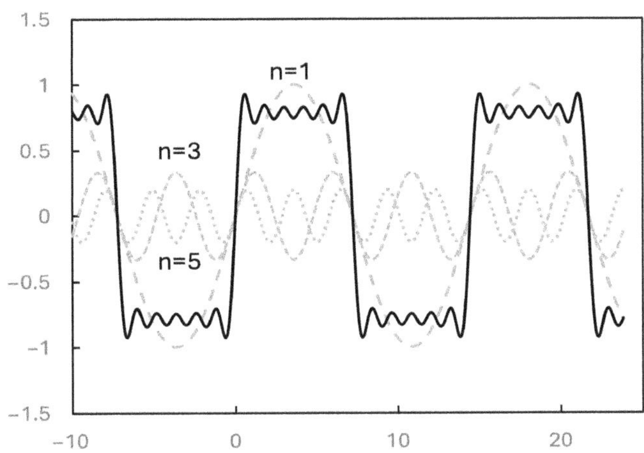

FIGURE 1.3
The first three terms of the Fourier representation of the periodic step function $f(x) = 1$ for $0 < x < L/2$ and $f(x) = -1$ for $L/2 < x < L$ (solid line) (for $L = 14.45$, x and L in arbitrary units); (dashed $n = 1$, light dashed $n = 3$, and dotted $n = 5$).

how they reflect properties under time reversal $t \to -t$. We proceed by discussing the damped harmonic oscillator as an illustrative example, cf. Fig. 1.4.

The invariance of Newton's laws under time reversal requires that forces are also invariant $F(t) = F(-t)$, which is the case for 'gravity' $F = mg$ and a linear restoring force (or spring force) $F = -kx = -m\Omega^2 x$. In contrast, damping describes irreversibility; for an object moving in a viscous fluid, the damping force is linear in the velocity $F_{\mathrm{d}} = -m\gamma v$ so that $F_{\mathrm{d}}(t) = -F_{\mathrm{d}}(-t)$. The equation of motion follows $\ddot{x} + \gamma\dot{x} + \Omega^2 x = 0$, where we adopt the compact notation $\dot{x} = dx/dt$ and $\ddot{x} = d^2x/dt^2$. In the *weak* damping limit $\gamma < \Omega$, we have an oscillatory solution $x(t) = Ae^{-\gamma t/2}\cos(\Omega' t + \phi)$, where $\Omega'^2 = \Omega^2 - \gamma^2/4$. The *strong* damping limit $\gamma > \Omega$ is not relevant for the content of this book.

We now consider the case of an external forcing, and first discuss a periodically-varying force

$$\ddot{x} + \gamma\dot{x} + \Omega^2 x = \frac{F_0}{m}\cos(\omega t). \tag{1.28}$$

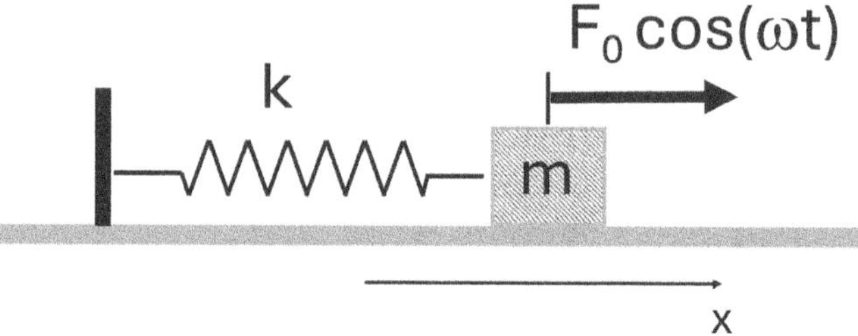

FIGURE 1.4
A block with mass m moving on a horizontal surface subject to a linear restoring (or spring) force $F = -kx = -m\Omega^2 x$ and viscous damping force $F_\text{d} = -m\gamma\dot{x}$. The block is subject to a periodically varying driving force $F_0 \cos(\omega t)$.

We find

$$x(t) = \frac{F_0}{m}\widehat{\chi}(\omega)\cos\left(\omega t + \phi(\omega)\right), \tag{1.29}$$

where

$$\widehat{\chi}(\omega) = \frac{1}{\sqrt{[\Omega^2 - \omega^2]^2 + \gamma^2\omega^2}}, \tag{1.30}$$

$$\phi(\omega) = \tan^{-1}\left(\frac{\gamma\omega}{\Omega^2 - \omega^2}\right). \tag{1.31}$$

The amplitude and phase of the susceptibility are shown in Fig. 1.5. Formally, the switch from time-*independent* to time-*dependent* displacements can be described by the change $1/k = 1/m\Omega^2 \rightarrow \chi(\omega)/m$.

In the limit $\omega/\Omega << 1$, the force is (essentially) constant during one cycle of the oscillator and the displacement follows the force $x \simeq F_0/k$ corresponding to $\widehat{\chi} = 1$; the block oscillates 'in synchrony' with the forcing which corresponds to a zero phase difference $\phi \rightarrow 0$. In the limit $\omega/\Omega >> 1$, the block is subject to a time-averaged force $F(t) \rightarrow \langle F(t)\rangle = (\Omega/2\pi)\int_t^{t+2\pi/\Omega} F(s)ds$ so that the amplitude decreases $\widehat{\chi} \sim (1/\omega)^2$. The acceleration of the block is 'in synchrony' with the average force which corresponds to a phase difference $\phi \simeq \pi$. The maximum displacement of block corresponds to case when the forcing is 'in synchrony' with the block's velocity the phase difference $\phi = \pi/2$ and the maximum power

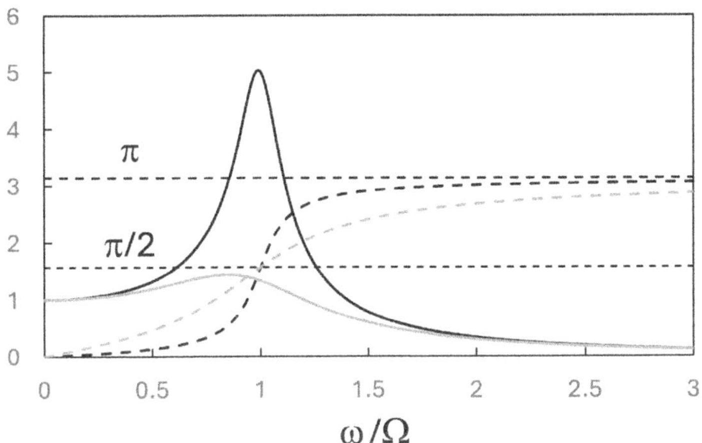

FIGURE 1.5
The amplitude χ (solid) and phase ϕ (dashed) for $\gamma/\omega = 0.2$ (dark gray) and $\gamma/\omega = 0.75$ (light gray).

'delivered' to damped harmonic oscillator by the external forcing. The frequency of the forcing matches the natural frequency $\omega = \Omega$ (so-called resonance) so that $\widehat{\chi} \simeq 1/\gamma\omega$.

The linear response is defined as the response to a delta-function forcing,

$$\frac{d^2}{dt^2}\chi(t-s) - \gamma\frac{d}{dt}\chi(t-s) + \Omega^2\chi(t-s) = \delta(t-s). \tag{1.32}$$

The dynamic susceptibility follows

$$\widehat{\chi}(\omega) = \frac{1}{\Omega^2 - \omega^2 - i\gamma\omega}; \tag{1.33}$$

The zeros of the denominator define the 'poles'

$$\widehat{\chi}(\omega) = \frac{1}{(\omega_+ - \omega_-)}\left[\frac{1}{(\omega-\omega_+)} - \frac{1}{(\omega-\omega_-)}\right]. \tag{1.34}$$

where the poles have real and imaginary parts $\omega_\pm = \pm\omega' + i\omega''$, with $\omega' = \sqrt{\Omega^2 - \gamma^2/4}$ and $\omega'' = \gamma/2$. We note that the two poles are located in the upper complex frequency plane and are symmetric along the real axis. Thus, the poles describe the frequencies at which the response is strong ('resonance' behavior). The real part describes oscillatory behavior and the imaginary part describes (exponential) decay $\exp(i\omega_{\pm}t) = e^{-\omega'' t}\cos(\pm\omega' t)$. Since damping leads to a decaying solution both in forward and backward directions, time reversal $t \to -t$ requires that the imaginary part of the frequency changes sign (whereas the real part remains the same);

$$t \to -t \quad \longleftrightarrow \quad \omega \to \omega^*, \tag{1.35}$$

that is, the complex-conjugate of the frequency.

The dynamic susceptibility is complex-valued and has a real and an imaginary part,

$$\widehat{\chi}(\omega) = \mathrm{Re}\,\widehat{\chi}(\omega) + i\,\mathrm{Im}\,\widehat{\chi}(\omega), \tag{1.36}$$

where $\mathrm{Re}\,\widehat{\chi}(\omega) = \int_0^\infty \cos(\omega t)\chi(t)dt$ and $\mathrm{Im}\,\widehat{\chi}'(\omega) = \int_0^\infty \sin(\omega t)\chi''$ $(t)dt$. The real and imaginary parts are even and odd functions, respectively, $\mathrm{Re}\,\widehat{\chi}(\omega) = \mathrm{Re}\,\widehat{\chi}(-\omega)$ and $\mathrm{Im}\,\widehat{\chi}(\omega) = -\mathrm{Im}\,\widehat{\chi}(-\omega)$. The dynamic susceptibility is an analytic function in the upper complex frequency plane. The relationship between analytic properties and the physical properties of the underlying physical properties are studied in graduate level texts on electrodynamics. In particular, the causality is captured in a relation between the real and imaginary parts of the dynamic susceptibility (so-called Kramers-Kronig relation).[2]

For small (large) frequencies, corresponding to the long- (short-) time behavior, the dynamic susceptibility has Taylor-series expansion in terms of the (angular) frequency $i\omega$: $\widehat{\chi}(\omega) = \widehat{\chi}(0) + \omega\,(d\widehat{\chi}/d\omega)|_0 + \frac{1}{2}(\omega)^2\,(d^2\widehat{\chi}/d\omega^2)\big|_0 + ...$ We examine the coefficients: (i) the zeroth-order term is $\widehat{\chi}(0) = \int_{-\infty}^{\infty}\chi(t)dt = \langle\chi(t)\rangle$, (ii) the first-order term is $(d\widehat{\chi}/d\omega)|_0 = (d/d\omega\int_{-\infty}^{\infty}e^{i\omega t}\chi(t)dt\Big|_0 = i\int_{-\infty}^{\infty}t\chi(t)dt = i\,\langle t\chi(t)\rangle$, and (iii) the second-order term is $(d^2\widehat{\chi}/d\omega^2)\big|_0 = (d^2/d\omega^2\int_{-\infty}^{\infty}e^{i\omega t}\chi(t)dt\Big|_0 = -\int_{-\infty}^{\infty}t^2\chi(t)dt = -\left\langle t^2\chi(t)\right\rangle$. We thus arrive at the expansion

$$\widehat{\chi}(\omega) = \langle\chi(t)\rangle + i\,\langle t\chi(t)\rangle\,\omega - \frac{1}{2}\left\langle t^2\chi(t)\right\rangle\omega^2 + ... \tag{1.37}$$

Writing the linear response as the sum of a symmetric and antisymmetric functions $\chi(t) = \chi_{\mathrm{s}}(t) + \chi_{\mathrm{a}}(t)$ with $\chi_{\mathrm{s}}(t) = [\chi(t) + \chi(-t)]/2$

[2]Causality and analytical properties of response functions are discussed in books on *Electrodynamics*, Jackson, Sect. 7.10.

and $\chi_a(t) = [\chi(t) - \chi(-t)]/2$, we have $\langle\chi(t)\rangle = \langle\chi_s(t)\rangle$, $\langle t\chi(t)\rangle = \langle t\chi_a(t)\rangle$ and $\langle t^2\chi(t)\rangle = \langle t^2\chi_s(t)\rangle$.

An analogous situation applies for the spatial dependence. The susceptibility

$$h(x) = \langle \chi * F_{\text{ext}} \rangle (x) = \int_0^x \chi(x - x')F_{\text{ext}}(x')dx' \qquad (1.38)$$

so that in reciprocal space,

$$\widehat{h}(\mathsf{k}) = \widehat{\chi}(\mathsf{k})\widehat{f}(\mathsf{k}). \qquad (1.39)$$

Here, $\mathsf{k} = 2\pi/\lambda$ is the wave vector. For the purpose of this book, we are only interested in matter that are nearly spatially homogeneous; that is, we seek an expansion for wavelengths much longer than typical molecular scale $\lambda >> a$. That is, we seek an expansion for small wavevectors $\mathsf{k} \to 0$. We now have the first-order correction to the spatial homogeneity,

$$\widehat{\chi}(\mathsf{k}) \simeq \langle\chi(x)\rangle + i\,\langle x\chi(x)\rangle\,\mathsf{k}. \qquad (1.40)$$

We write the spatial response in terms of a symmetric and antisymmetric contribution $\chi(x) = \chi_s(x) + \chi_a(x)$ with $\chi_s(x) = [\chi(x) + \chi(-x)]/2$ and $\chi_a(x) = [\chi(x) - \chi(-x)]/2$. Thus, the first-order correction vanishes in systems with inversion symmetry $\chi(x) = \chi(-x)$ [or $\chi(\vec{r}) = \chi(-\vec{r})$ in two or thee dimensions]. We write $l = \langle x\chi_a(x)\rangle$ and find that the first-order correction describes spatial decay in systems *without* spatial inversion $\chi(x) = \exp(-x/l)$. That is, it describes a decaying amplitude and thus mimic role of viscous forces in the time- (or frequency)-dependent systems. The latter is dominant in systems with and without spatial inversion so that spatial inversion is usually ignored.

We seek a susceptibility for which the linear term is real and thus describes a wave-like, rather than dissipative, behavior. This requires that the wave vector is imaginary $\mathsf{k} = i\mathsf{k}''$, which is unphysical in one spatial dimension. In three dimensions, however, multiplication with a complex number corresponds to a rotation. More specifically, an imaginary wave vector $\vec{\mathsf{k}} = i\mathsf{k}_z\hat{z}$ describes a rotation in the (x, y) plane. We conclude that the linear term involves a third-order tensor γ_{lmn}, where we adopt the notation $(x = x_1,\ y = x_2,\ z = x_3)$. We write

$$\widehat{\chi}_{lm}(\mathsf{k}) = \widehat{\chi}_{lm}(0)\,\delta_{lm} + i\sum_n \gamma_{lmn}\,\mathsf{k}_n. \qquad (1.41)$$

Since multiplication with a wave vector corresponds to taking the derivative with respect to a coordinate, $i\mathsf{k}_n \leftrightarrow \partial/\partial x_n$, we have

$$\widehat{h}_l = \sum_m \widehat{\chi}_{lm}(0)\, F_{\text{ext},m} + \sum_{mn} \gamma_{lmn} \frac{\partial F_{\text{ext},m}}{\partial x_n}.$$

If the third-ranked tensor is proportional to the antisymmetric tensor (Levi-Civita symbol) $\gamma_{lmn} = \gamma\epsilon_{lmn}$, the linear term is proportional to the curl of the force

$$\sum_{mn} \gamma_{lmn} \frac{\partial F_{\text{ext},m}}{\partial x_n} \quad \longleftrightarrow \quad \gamma\nabla \times \vec{F}_{\text{ext}}. \tag{1.42}$$

1.2 Units of Physical Quantities

Mechanical systems: Reading older research texts is often made challenging by their use of units, rather unfamiliar to a modern reader. We assume that the reader is familiar with the MKS (SI) system of units in which mass is measured in kilogram $[m] = \text{kg}$, length in meter $[L] = \text{m}$, and time in seconds $[t] = \text{s}$. Up until the 1960s, the cgs (Gaussian) system was used in which mass is measured in gram $[m] = \text{g}$, length in centimeter $[L] = \text{cm}$, and time in seconds $[t] = \text{s}$.

The units of other mechanical quantities are derived. For forces, we have the unit of *newtons*, $[F] = \text{N} = \text{kg}\cdot\text{m}/\text{s}^2$. In the cgs system we have the unit of *dynes*,

$$[F] = \text{dyn} = \text{g} \cdot \frac{\text{cm}}{\text{s}^2}. \tag{1.43}$$

We have the conversion

$$1\,\text{N} = 10^5\,\text{dyn}. \tag{1.44}$$

The elastic properties of a spring define the spring constant k with unit $[k] = \text{N}/\text{m} = \text{kg}/\text{s}^2$ in the MKS system, and in the cgs system

$$[k] = \frac{\text{dyn}}{\text{cm}} = \frac{\text{g}}{\text{s}^2} = 10^{-3}\,\frac{\text{N}}{\text{m}}. \tag{1.45}$$

For energies, we have the unit of *joules* in the MKS system, $[U] = \text{J} = \text{kg} \cdot \text{m}^2/\text{s}^2$. In the cgs system, we have the unit of 'erg,'

$$[U] = \text{erg} = \text{g} \cdot \frac{\text{cm}^2}{\text{s}^2}. \tag{1.46}$$

We have the conversion

$$1\,\mathrm{J} = 10^7\,\mathrm{erg}. \tag{1.47}$$

The conversion between MKS and cgs units is straightforward for mechanical problem and only involves power of 10.

Electricity and magnetism: The situation is more challenging for problems in electricity and magnetism. In the SI system, the unit of *ampere* for the electric current $[I] = \mathrm{A}$ is defined by the force between two parallel wires. The unit of electric charge *coulomb* is a derived unit $[Q] = \mathrm{C} = \mathrm{A} \cdot \mathrm{s}$; 1 coulomb is an enormous charge (about the order of the charge delivered in a lightning strike). The magnitude of the (Coulomb-) force between two point charges separated by the distance r in free space is written as

$$F = \frac{1}{4\pi\epsilon_0}\frac{|q_1| \cdot |q_2|}{r^2} = k\frac{|q_1| \cdot |q_2|}{r^2}, \tag{1.48}$$

where $\epsilon_0 = 8.85 \times 10^{-12}\,\mathrm{C^2 s^2/(kg\,m^3)}$ is the electric permeability of free space and $k = 8.99 \times 10^9\,\mathrm{kg} \cdot \mathrm{m^3/(C^2 \cdot s^2)}$ is the Coulomb constant.

In the cgs system, the Coulomb constant is set equal to unity and Coulomb's law is written as

$$F = \frac{|q_1| \cdot |q_2|}{r^2}. \tag{1.49}$$

The statcoulomb (statC) is defined by setting $F = 1\,\mathrm{dyn}$ and $r = 1\,\mathrm{cm}$

$$\mathrm{statC} = \mathrm{dyn}^{1/2} \cdot \mathrm{cm} = \frac{\mathrm{g}^{1/2} \cdot \mathrm{cm}^{3/2}}{\mathrm{s}}. \tag{1.50}$$

We set $F = 10^{-5}\,\mathrm{N}$ and $r = 10^{-2}\,\mathrm{m}$; it follows that 1 statcoulomb corresponds to a charge $q \simeq (1/3) \times 10^{-9}\,\mathrm{C}$. Electric charges in the MKS and cgs systems cannot be compared to each other since the dimensions; this applies not only to the electric charge but also derived quantities such as the electric field or electric potential. With this caveat in mind, we write

$$1\,\mathrm{C} = 2.998 \times 10^9\,\mathrm{statC}. \tag{1.51}$$

The electric field is defined as force per charge $E = F/q$; this definition implies that the electric field vector points from regions with (predominantly) positive charges to regions with predominantly negative charges. Because the electric field vector is defined via the position of charges, it follows that the electric field vector is a *polar* vector under inversion

$$\vec{E} \to \vec{E'} = -\vec{E} \quad \text{for} \quad \vec{r} \to \vec{r'} = -\vec{r}. \tag{1.52}$$

The unit in the MKS system is $[E] = \text{N/C} = (\text{kg} \cdot \text{m})/(\text{C} \cdot \text{s}^2)$ and in the cgs system, $[E] = \text{dyn/statC} = \text{g}^{1/2}/(\text{cm}^{1/2} \cdot \text{s})$ in the cgs system. The energy density follows

$$u_E = \frac{\epsilon}{2} E^2 \quad \text{(MKS)} \qquad u_E = \frac{1}{8\pi} E^2 \quad \text{(cgs)}; \tag{1.53}$$

we check the unit in MKS system $[u_E] = \text{J/m}^3 = \text{kg}/(\text{m} \cdot \text{s}^2)$ and in the cgs system $[u_E] = \text{erg/cm}^3 = \text{g}/(\text{cm} \cdot \text{s}^2)$. The electrostatic potential Φ is defined as energy per charge with unit of *volts* $[\Phi] = \text{J/C} = \text{V}$ in the SI system and in the cgs system, we define the *statvolt*

$$[\Phi] = \text{statV} = \frac{\text{erg}}{\text{statC}} = \frac{\text{g}^{1/2} \cdot \text{cm}^{1/2}}{\text{s}}. \tag{1.54}$$

We have the conversion between volts and statvolts, $1\,\text{V} = (10^7\,\text{erg})/(2.998 \times 10^9\,\text{statC})$ so that

$$1\,\text{V} = \frac{1}{299.8}\,\text{statV}. \tag{1.55}$$

A static electric field is derived from the electrostatic potential $\vec{E} = -\nabla\Phi$. The unit of electric field follows $[E] = \text{V/m}$ in the MKS system and $[E] = \text{statV/cm}$ in the cgs system, $[E] = \text{statV/cm}$. We have

$$1\,\frac{\text{V}}{\text{m}} = \frac{1}{2.998 \times 10^4}\,\frac{\text{statV}}{\text{cm}}. \tag{1.56}$$

A positive $q_+ = +q_0$ and negative charge $q_- = -q_0$ separated by the distance a form an electric dipole $p = q_0 a$; the unit is $[p] = \text{C}\cdot\text{m}$ (in MKS) and $[p] = \text{statC}\cdot\text{cm}$ ($\text{C}\cdot\text{m} = 2.998 \times 10^7\,\text{statC}\cdot\text{cm}$). The direction of the dipole moment is defined from the negative to the positive charge. In the presence of an external electric field $\vec{E}$, the potential energy is given by

$$\text{PE} = -\vec{p} \cdot \vec{E}, \tag{1.57}$$

in both the MKS and cgs systems, which implies that the dipole orients itself *parallel* to the electric field. The net charge of a dipole is zero so that the an electric dipole at $\vec{r}_0$ creates an electric field that decreases faster than the inverse-square-law,

$$\vec{E}(\vec{r}) = \frac{1}{4\pi\epsilon_0}\frac{3\hat{n}(\vec{p}\cdot\hat{n}) - \vec{p}}{|\vec{r}-\vec{r}_0|^3} \text{ (MKS)}, \quad \vec{E}(\vec{r}) = \frac{3\hat{n}(\vec{p}\cdot\hat{n}) - \vec{p}}{|\vec{r}-\vec{r}_0|^3} \text{ (cgs)}. \tag{1.58}$$

Here, $\hat{n} = (\vec{r}-\vec{r}_0)/|\vec{r}-\vec{r}_0|$. Eqs. (1.57) and (1.58) imply that two electric dipoles interact with each other

$$\mathrm{PE_{dd}} = \frac{1}{4\pi\epsilon_0}\frac{\vec{p}_1\cdot\vec{p}_2 - 3(\hat{n}\cdot\vec{p}_1)(\hat{n}\cdot\vec{p}_2)}{|\vec{r}-\vec{r}_0|^3} \text{ (MKS)}, \tag{1.59}$$

$$\mathrm{PE_{dd}} = \frac{\vec{p}_1\cdot\vec{p}_2 - 3(\hat{n}\cdot\vec{p}_1)(\hat{n}\cdot\vec{p}_2)}{|\vec{r}-\vec{r}_0|^3} \text{ (cgs)}.$$

A net charge (or *monopole* moment) and the dipole moment are the two lowest terms of an expansion of the electric potential of a charge distribution in inverse power of the radius r (so-called multipole expansion).

For magnetic systems, we start from the Lorentz force on a moving charge. In the MKS system, the force is written as

$$\vec{F} = q(\vec{E} + \vec{v}\times\vec{B}) \text{ (MKS)} \quad \vec{F} = q\left(\vec{E} + \frac{1}{c^2}\vec{v}\times\vec{B}\right) \text{ (cgs)}, \tag{1.60}$$

where $c = 2.998\times 10^{10}$ cm/s is the speed of light (in free space). In the MKS system, the electric and magnetic fields have different units. The unit for magnetic field is *tesla* $[B] = [E]/[v] = \mathrm{T} = \mathrm{kg}/(\mathrm{C}\cdot\mathrm{s})$. In the cgs system, the electric and magnetic fields have the same unit $[E] = [B] = \mathrm{g}^{1/2}/(\mathrm{cm}^{1/2}\cdot\mathrm{s})$. Since $1\,\mathrm{T} = 1\,(\mathrm{V}\cdot\mathrm{s})/\mathrm{m}^2$, we find

$$1\,\mathrm{T} = \frac{1}{2.998\times 10^6}\frac{\mathrm{statV}\cdot\mathrm{s}}{\mathrm{cm}^2}. \tag{1.61}$$

We have under an inversion ($\vec{r}\to\vec{r'} = -\vec{r}$), $\vec{F}\to\vec{F'} = -\vec{F}$ and $\vec{v}\to\vec{v'} = -\vec{v}$, we conclude that the magnetic must be an axial vector

$$\vec{B}\to\vec{B'} = \vec{B} \quad \text{for} \quad \vec{r}\to\vec{r'} = -\vec{r}. \tag{1.62}$$

The magnetic field produced by a long thin wire is given by $\vec{B} = (\mu_0 I/2\pi r)\hat{\imath}\times\hat{r}$ (MKS) and $\vec{B} = (2I/rc)\hat{\imath}\times\hat{r}$ (cgs), where the magnetic permeability of free space is $\mu_0 = 4\pi\times 10^{-7}\,\mathrm{T}\cdot\mathrm{m/A}$ and $\hat{\imath}$ is the unit vector along the wire in the direction of the current (defined as the flow of *positive* charges) and $\hat{r} = \vec{r}/r$. In the cgs system, the magnitude of the magnetic field follows $B = 2I/(rc)$.

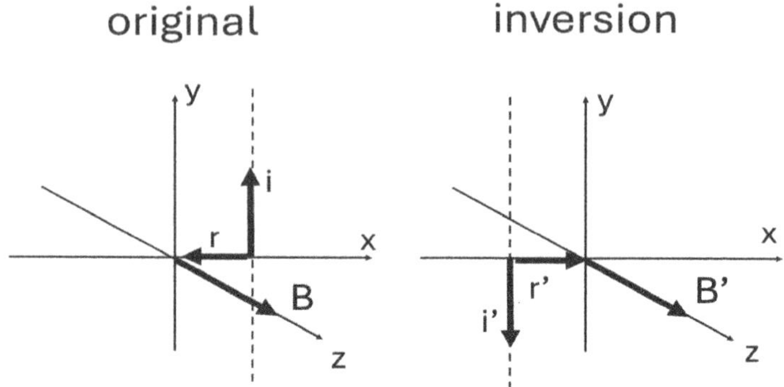

FIGURE 1.6
In the original version, the current flows along the $+y$-axis at a point on the $+x$-axis. In the inverted version, the current flows along the $-y$-axis at a point on the $-x$-axis. The magnetic field $\vec{B}$ at the origin point along the $+z$-axis.

The axial character of the magnetic field is illustrated in Fig. 1.6. We check the units $[B] = (\text{statC/s})/(\text{cm} \cdot \text{cm/s}) = (\text{g}^{1/2} \cdot \text{cm}^{3/2} \cdot \text{s}^{-2})/(\text{cm}^2 \cdot \text{s}^{-1}) = (\text{g})/(\text{cm}^{1/2} \cdot \text{s})$; thus electric and magnetic fields have the same units $[E] = [B]$, as it should.

We consider a long thin wire with a current in the $+x$-direction located along the $+y$ direction. The magnetic field at the origin is directed along the $-z$-axis. The inversion yields a current flowing in the $-x$-direction with the wire located along the $-y$-axis. The magnetic field does not change, and we recover that the magnetic field is described by an axial vector. The energy density of a magnetic field is

$$u_B = \frac{B^2}{2\mu_0} \quad (\text{MKS}) \qquad u_B = \frac{1}{8\pi} B^2 \quad (\text{cgs}). \tag{1.63}$$

1.3 Electrodynamics

In vacuum, electrodynamic phenomena are described by electric and magnetic fields. These fields cannot be separated from each other by the principles of special relativity; this connection is irrelevant for the purpose here. In the static case (i.e., non-moving charges), electric charges produce an electric field $\vec{E}(\vec{r})$. The corresponding electric field lines start at positive and end at negative charges. Similarly, stationary currents produce magnetic fields $\vec{B}(\vec{r})$. The corresponding magnetic field lines have no beginning or end; we say that the magnetic field line 'curl' around currents. In introductory texts, the orientation (right- or left-circular) is usually referred to as 'first-right-hand rule.' This simple description has to be modified in two directions: we first discuss the role of matter and then we summarize time-dependent phenomena.

The basic properties of electric fields in the presence of a dielectric material are usually discussed in the simple case of a charged capacitor. Without dielectric material, the surface charge on the plates (so-called 'free' surface charge) is $\sigma_{\text{free}} = Q/A$. The electric field inside the capacitor follow $E = \sigma_{\text{free}}/\epsilon_0$ (in the MKS system) and $E = 4\pi\sigma_{\text{free}}$ (in the cgs system). We assume that the lower (upper) plate is at the higher (lower) potential and thus is positively (negatively) charged; it follows the electric field is directed *upwards.*

We insert a dielectric slab inside the capacitor; the lower (upper) surface carries a negative (positive) charge density $\sigma_{\text{pol}} = P$; that is, σ_{free} and σ_{pol} have opposite signs. The electric field inside the dielectric is weakened: $E = (\sigma_{\text{free}} - \sigma_{\text{pol}})/\epsilon_0 = (\sigma_{\text{free}} - P)/\epsilon_0$ (in the MKS system) and $E = 4\pi(\sigma_{\text{free}} - \sigma_{\text{pol}}) = 4\pi(\sigma_{\text{free}} - P)$ (in the cgs system).

We recall the dipole moment $\vec{p}$ is defined as directed from the negative to the positive charge; thus, the polarization vector is also directed *upwards.* The electric polarization is defined as dipole moment per volume $\vec{P} = \vec{p}/\text{volume}$. In the simplest case, the polarization is proportional to the electric field

$$\vec{P} = \chi_E \epsilon_0 \vec{E} \quad (\text{MKS}) \qquad \vec{P} = \chi_E \vec{E} \quad (\text{cgs}), \tag{1.64}$$

here, χ_E is the electric susceptibility. We define the electric *displacement,*

$$\vec{D} = \epsilon_0 \vec{E} + \vec{P} \quad (\text{MKS}) \qquad \vec{D} = \vec{E} + 4\pi\vec{P} \quad (\text{cgs}). \tag{1.65}$$

The electric displacement is determined by free charges The relationship between the fields D and E defines the electric permittivity, or relative dielectric constant

$$\vec{D} = \epsilon\vec{E} \quad (\text{MKS}) \qquad \vec{D} = \kappa\vec{E}, \quad (\text{cgs}), \tag{1.66}$$

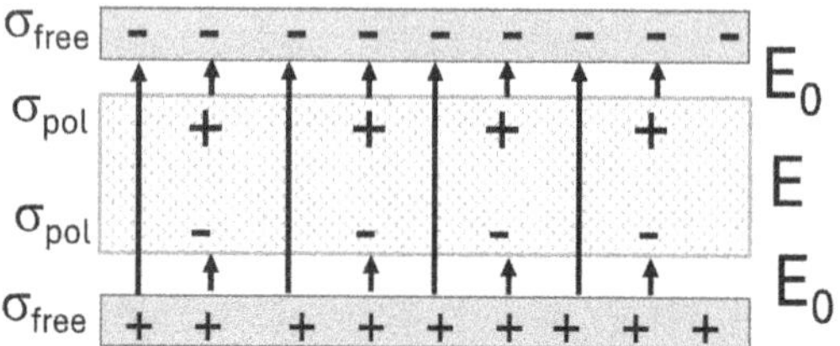

FIGURE 1.7
A dielectric inside a parallel-plate capacitor. The charge densities on the top and bottom plates are $\sigma_{\text{top}} = -\sigma_{\text{free}}$ and $\sigma_{\text{bottom}} = +\sigma_{\text{free}}$ so the electric fields point from the bottom to the top plate. The induced charge densities at the surface of the dielectric have opposite polarity: $\sigma'_{\text{top}} = +\sigma_{\text{pol}}$ and $\sigma'_{\text{bottom}} = -\sigma_{\text{pol}}$ so the electric fields point from the bottom to the top plate.

where

$$\epsilon = (1+\chi)\epsilon_0 = \epsilon_{\text{rel}}\epsilon_0 \quad \text{(MKS)} \qquad \kappa = 1 + 4\pi\chi \quad \text{(cgs)}. \tag{1.67}$$

For a spatially homogenous field, the surface polarization of a dielectric slab is shown in Fig. 1.7.

The magnetic moment is defined as an elementary current loop enclosing the area $|\vec{a}|$,

$$\vec{m} = I\vec{a} \quad \text{(MKS)} \qquad \vec{m} = \frac{1}{c}(I\vec{a}) \quad \text{(cgs)}. \tag{1.68}$$

We analyze the case of single charge with charge q and mass M moving along a circle with radius a with period T. This corresponds to the current $I = q/T$ so that the magnetic moment follows $m = (q/T)\pi a^2 = (q/2)\omega a^2$. Since ma^2 is the moment of inertia and $Ma^2\omega$ is the angular momentum, it follows that the magnetic moment is proportional to the angular moment

$$m = \gamma L \quad \text{with} \quad \gamma = \frac{q}{2M}, \tag{1.69}$$

where γ is referred to as gyromagnetic ratio. A magnetic moment placed in a region with magnetic field has the potential energy

$$U = -\vec{m} \cdot \vec{B}. \tag{1.70}$$

Since $\nabla \cdot \vec{B} = 0$, the force on a magnetic moment is given by the negative gradient, $\vec{F} = \nabla(\vec{m} \cdot \vec{B})$.

The magnetization is defined as magnetic moment per unit volume $\vec{M} = \vec{m}/\text{volume}$ so that the current volume density follows

$$\vec{j} = \nabla \times \vec{M} \quad \text{(MKS)} \qquad \vec{j} = c\nabla \times \vec{M} \quad \text{(cgs)}. \tag{1.71}$$

We consider the case when the magnetization is due to a magnetic field,

$$\vec{M} = \frac{\chi_M}{\mu_0}\vec{B} \quad \text{(MKS)} \qquad \vec{M} = \chi_M \vec{B} \quad \text{(cgs)}. \tag{1.72}$$

The *auxiliary* field $\vec{H}$ is defined as

$$\vec{H} = \frac{\vec{B}}{\mu_0} - \vec{M} \quad \text{(MKS)} \qquad \vec{H} = \vec{B} - 4\pi\vec{M} \quad \text{(cgs)}. \tag{1.73}$$

The magnetic permeability is defined as

$$\mu = \mu_0(1 + \chi_M) \quad \text{(MKS)} \qquad \mu = 1 + 4\pi\chi_M \quad \text{(cgs)}. \tag{1.74}$$

We then have $\vec{B} = \mu\vec{H}$ in both the MKS and cgs sytems.

The connection between electric and magnetic phenomena is caused by the time-dependence of electromagnetic fields: a time-dependent magnetic field (described, for example, by moving a permanent magnet toward a wire loop) causes an *EMF* (*E*lectro*M*otive *F*orce); that is, electric field lines with no beginning or end. The orientation of electric files (right- or left-handed) is captured by Lenz' law: the magnetic field produced by induced currents oppose the change. Similarly, a time-dependent electric field is equivalent to a current density (the so-called Maxwell displacement current); in introductory texts, this is usually discussed in the context of (dis-)charging a parallel-plate capacitor.

The fundamental fields are $\vec{E}$ and $\vec{B}$ and the macroscopic fields are $\vec{D}$ and $\vec{H}$. Maxwell's equations for dielectric matter then follow

$$\nabla \cdot \vec{D} = \rho_{\text{free}} \quad \text{(MKS)} \qquad \nabla \cdot \vec{E} = 4\pi\rho_{\text{free}} \quad \text{(cgs)}, \tag{1.75}$$

$$\nabla \times \vec{H} - \frac{\partial \vec{D}}{\partial t} = \vec{j} \quad \text{(MKS)} \qquad \nabla \times \vec{H} - \frac{1}{c}\frac{\partial \vec{D}}{\partial t} = \frac{4\pi}{c}\vec{j}_{\text{free}} \quad \text{(cgs)}, \tag{1.76}$$

$$\nabla \times \vec{E} + \frac{\partial \vec{B}}{\partial t} = 0 \quad \text{(MKS)} \qquad \nabla \times \vec{E} + \frac{1}{c}\frac{\partial \vec{B}}{\partial t} = 0 \quad \text{(cgs)}, \tag{1.77}$$

$$\nabla \cdot \vec{B} = 0 \quad \text{(MKS)} \qquad \nabla \cdot \vec{B} = 0 \quad \text{(cgs)}. \tag{1.78}$$

We examine the properties under spatial inversion $\vec{r} \to \vec{r}' = -\vec{r}$ so that $\nabla \to \nabla' = -\nabla$. The electric field and the electric displacement are polar vectors and change signs $\vec{E} \to \vec{E}' = -\vec{E}$ and $\vec{D} \to \vec{D}' = -\vec{D}$. The magnetic fields are axial vectors and thus are invariant $\vec{B} \to \vec{B}' = \vec{B}$ and $\vec{H} \to \vec{H}' = \vec{H}$. We thus have $\nabla \times \vec{H} \to \nabla' \times \vec{H}' = -\nabla \times \vec{H}$ and $\partial\vec{D}/\partial t \to \partial\vec{D}'/\partial t = -\partial\vec{D}/\partial t$ and similarly $\nabla \times \vec{E} \to \nabla' \times \vec{E}' = \nabla \times \vec{E}$ and $\partial\vec{B}/\partial t \to \partial\vec{B}'/\partial t = \partial\vec{B}/\partial t$. We conclude that in the absence of external sources (charges and currents), Maxwell's equations are invariant under spatial inversion; so that, in particular, the solutions can always be chosen such that they are symmetric under spatial inversion.

Dielectric properties of matter describe the response of polarizable matter to an external electric field; this is analogous to the response of a mechanical system to an external force with the electric field corresponding to the force and the polarization corresponding to the coordinate displacement. As such, it is convenient to treat the polarization and the electric field as complex quantities (with the convention that the real part represents the physically relevant values). It follows that the dielectric response is complex valued (similar conclusions apply to the magnetic response but is not discussed here).

So far, we assumed that the dielectric response (and thus the (relative) electric permeability is a scalar: an electric field along the x-axis, say, produces a polarization along the x-axis. In a crystal, an electric field E_x can produce a polarization along all three coordinate axes: $D_x = \epsilon_{xx}E_x$, $D_y = \epsilon_{yx}E_x$, and $D_z = \epsilon_{zx}E_x$ (and similar for electric fields along the two other coordinate axes E_y and E_z). Such off-diagonal matrix elements are possible due to the shear strain in crystals.

The imaginary part of the electric permeability describes dissipation (damping) while the real part characterizes the systematic (non-dissipative) response; that is, the absence of dissipation implies that the imaginary part is zero, $\text{Im}\chi = \chi - \chi^* = 0$. This condition is generalized for the complex-valued electric susceptibility tensor

$$\epsilon_{lm} = \epsilon^*_{ml}; \tag{1.79}$$

that is, the tensor is *Hermetian.* It follows that real (imaginary) part is symmetric (antisymmetric):

$$\epsilon'_{lm} = \epsilon'_{ml} \qquad \epsilon''_{lm} = -\epsilon''_{ml}. \tag{1.80}$$

The (di-)electric properties of matter are determined by both atoms (ions) and (valence-) electrons. The motion of atoms (ions) undergoes oscillations (vibrations) around (mechanical) equilibrium positions, and can be described by the laws of classical mechanics [Newton's laws]. Electrons require a quantum-mechanical description and their dynamics does not define a characteristic time scale; for dielectric properties of matter, electrons can be described in terms of their probability distribution (i.e., their 'clouds'). The 'geometry' of electron clouds are determined by the location of atoms (ions); thus, the dielectric permeability depends on the distribution of atoms (ions) $\{\vec{r}_n\}$: $\epsilon_{lm} = \epsilon_{lm}(\{\vec{r}\}_n)$.

So far, we assumed that the dielectric response (and thus the [relative] electric permeability) is a scalar: an electric field along the x-axis, say, produces a polarization along the x-axis. In general, the electric field E_x can produce a polarization along all three coordinate axes: $D_x = \epsilon_{xx} E_x$, $D_y = \epsilon_{yx} E_x$, and $D_z = \epsilon_{zx} E_x$ (and similar for electric fields along the two other coordinate axes E_y and E_z).

The quantities (ϵ_{xx}, ϵ_{xy},...) are the elements of a (second-rank) tensor; they are examples of so-called kinetic coefficients discussed in the context of linear response theory. Properties of the coefficients follow general principles ('regression hypothesis' and fluctuations-dissipation theorem). Onsager derived symmetry properties by examining the local entropy production: irreversibility requires an external magnetic field must be 'flipped' $\vec{B} \rightarrow -\vec{B}$:

$$\epsilon_{lm}(\vec{B}) = \epsilon_{ml}(-\vec{B}) \quad l, m = x, y, z. \tag{1.81}$$

The imaginary part of the dynamic susceptibility describes dissipation (damping) while the real part characterizes the systematic (non-dissipative) response; that is, the absence of dissipation implies that the imaginary part is zero, $Im\chi = \chi - \chi^* = 0$. This condition is generalized for the complex-valued electric susceptibility tensor

$$\epsilon_{lm} = \epsilon^*_{ml}; \tag{1.82}$$

that is, the tensor is *Hermetian*. It follows that real (imaginary) part is symmetric (antisymmetric):

$$\epsilon'_{lm} = \epsilon'_{ml} \quad \epsilon''_{lm} = -\epsilon''_{ml}. \tag{1.83}$$

Thus, we have in the presence of a magnetic field

$$\epsilon'_{lm}(\vec{B}) = \epsilon'_{ml}(\vec{B}) = \epsilon'_{lm}(-\vec{B}) \quad \epsilon''_{lm}(\vec{B}) = -\epsilon''_{ml}(\vec{B}) = -\epsilon''_{lm}(-\vec{B}), \tag{1.84}$$

that is, the real (imaginary) part of the dielectric tensor is an even (odd) function of the magnetic field. We define the inverse tensor $\eta = \epsilon^{-1}$; the properties of the tensor elements η_{lm}, $(\epsilon^{-1})_{lm} = \eta_{lm} = \eta'_{lm} + i\eta''_{lm}$ are the same as those of ϵ_{lm}.

Electromagnetic potentials: In electrostatics, the motivation for introducing the electrostatic potential is the *conservative* nature of the electric force; that is, the work done by the electric force only depends on the start and end points but is independent of the path. Evidently, the work is zero when the endpoint is identical with the start point so that the work done over a closed loop vanishes $\oint \vec{E} \cdot d\vec{r} = 0$. This is mathematically equivalent with the statement that the electric field can be derived from a potential $\vec{E} = -\nabla\phi$. Here the negative sign is a convention: it is the same convention as in introductory physics where the work done by a conservative is equal to the negative change of the potential energy.

The idea of potential can also be expanded to the case of time-dependent electric and magnetic fields. The condition $\nabla \cdot \vec{B}$ requires that the magnetic field is derived from a vector potential $\vec{A}$ (in MKS unit) $\vec{B} = \nabla \times \vec{A}$ and $\vec{E} = -\nabla\phi + \partial\vec{A}/\partial t$, where the potentials are space- and time-dependent $\phi = \phi(\vec{r}, t)$ and $\vec{A} = \vec{A}(\vec{r}, t)$.

The vector potential and scalar potential are not uniquely defined: $\vec{A} \to \vec{A} + \nabla\Lambda$ and $\phi \to \phi' = \phi - \partial\Lambda/\partial t$. If the potentials satisfy the Lorentz condition $\nabla \cdot \vec{A} + c^{-2}\partial\phi/\partial t = 0$, one finds inhomogeneous wave equation $\nabla^2\phi - c^{-2}\partial^2\phi/\partial t^2 = -\rho/\epsilon_0$ and $\nabla^2\vec{A} - c^{-2}\partial^2\vec{A}/\partial t^2 = -\mu_0\vec{j}$. In the Lorentz gauge, $\nabla \cdot \vec{A} = 0$, the scalar potential satisfy the Poisson equation $\nabla^2\phi = -\rho/\epsilon_0$; that is, the scalar potential depends on the instantaneous charge distribution. Thus, in the region without sources $\rho = 0$ and $\vec{j} = 0$, the scalar potential vanishes $\phi = 0$ and the electric and magnetic fields are determined by the vector potential

$$\vec{E} = -\frac{\partial\vec{A}}{\partial t} \tag{1.85}$$

$$\vec{B} = \nabla \times \vec{A}, \tag{1.86}$$

where the vector potential satisfies the homogenous wave equation.

Energy, momentum, and angular momentum: Electromagnetic fields carry energy and momentum. The intensity (power per area with unit W/m^2); it is given by the Poynting vector

$\vec{S} = \vec{E} \times \vec{H}$, or (in vacuum)

$$\vec{S} = \frac{1}{\mu_0}\vec{E} \times \vec{B} \quad \text{(MKS)} \quad \vec{S} = \frac{c}{4\pi}\vec{E} \times \vec{B} \quad \text{(cgs)}, \tag{1.87}$$

or in terms of the electric field, $S = c\epsilon_0 \langle E^2 \rangle$ in the (MKS) and $S = (c/4\pi)\langle E^2 \rangle$ (cgs). The units are $[S] = \mathrm{J/(m^2 \cdot s)}$ (MKS) and $[S] = \mathrm{erg/(cm^2 \cdot s)}$ (cgs). We consider numerical values and set $S = 1\,\mathrm{W/m}^2 = 10^3\,\mathrm{erg/(s\,cm^2)}$. The energy density follows $u = 3{\times}10^{-9}\,\mathrm{J/m}^3 = 3{\times}10^{-8}\,\mathrm{erg/cm}^3$, and the corresponding (rms) electric field $E_{\mathrm{rms}} \simeq 20\,\mathrm{V/m} \simeq 7 \times 10^{-4}\,\mathrm{statV/cm}$. We consider numerical values and set $S = 1\,\mathrm{kW/m}^2$. This corresponds to a electric field $E_{\mathrm{rms}} \simeq 200\,\mathrm{V/m}$ (or $6.0 \times 10^{-5}\,\mathrm{statV/cm}$).

The momentum density follows $\vec{g} = \vec{S}/c^2$, or

$$\vec{g} = \epsilon_0\vec{E} \times \vec{B} \quad \text{(MKS)} \quad \vec{g} = \frac{1}{4\pi c}\vec{E} \times \vec{B}. \tag{1.88}$$

with unit $\mathrm{kg/(m^2 s)}$. The angular momentum density follows

$$\vec{L} = \frac{1}{\mu_0 c^2}\vec{r}{\times}(\vec{E}{\times}\vec{B}) \quad \text{(MKS)} \quad \vec{L} = \frac{1}{4\pi c}\vec{r}{\times}(\vec{E}{\times}\vec{B}) \quad \text{(cgs)}. \tag{1.89}$$

Alternatively, the angular momentum can be written in terms of the vector potential $\vec{A}$.[3] The relationship between energy and momentum fluxes are more familiar in the photon description. of light. For an electromagnetic wave with frequency f and wavelength λ, the energy and momentum of a photon are $E_\gamma = hf$ and $p_\gamma = h/\lambda$ related to the energy-momentum relation of particle with zero mass $m_\gamma = 0$: $E_\gamma = p_\gamma c$. The angular momentum relates to the helicity of photons: left- (right-)circularly polarized light corresponds to photons with positive (negative) helicity $L = +\hbar$ ($L = -\hbar$).

The energy dissipation due to a monochromatic electromagnetic field follows for nonmagnetic system ($\mu_{ik} = \mu_0$): $Q = i\omega\epsilon_0(\epsilon^*_{ij} - \epsilon_{ji})E_i E^*_j$ (MKS) $Q = (i\omega/16\pi)(\epsilon^*_{ij} - \epsilon_{ji})E_i E^*_j$ (cgs) where we use a complex notation for the electric fields. Thus, dissipation of electromagnetic waves is associated with a non-Hermetian character of the dielectric tensor.

[3]The angular moment follows $\vec{L} = (\mu_0 c^2)^{-1}\left[\vec{E} \times \vec{A} + \sum_n E_n(\vec{r} \times \nabla)A_n\right]$.

1.4 Electromagnetic Waves

For an isotropic linear dielectric medium, we have $\vec{D} = \epsilon_r \epsilon_0 \vec{E}$. We assume that there are no free charges and currents. We proceed with the macroscopic description of electric and magnetic fields in the MKS system. The existence of a propagating electromagnetic wave is made possible by the *time-dependent* coupling between electric field and displacement $\vec{E}$ and $\vec{D}$ and magnetic field $\vec{B}$ that stems from Faraday's laws, cf, Eq. (1.73) and Maxwell's displacement current, cf. Eq. (1.72). Additionally, we need a material equation that relates the electric displacement to the electric field.

In a 'typical' experiment, a slab (with thickness d) of dielectric material is inserted into a beam of (monochromatic) light, and its effect on the propagating electromagnetic wave is studied. That is, we consider $(\vec{E}, \vec{B})$ as the primary field; we seek to find a *qualitative* description of the effect of dielectric matter on the propagation of these fields. To this end, we discuss the case of a fixed value of the relative dielectric constant $\epsilon_r > 0$. We introduce scaled units: we measure time in length $t = \tau/c$ and measure the magnetic field in terms of an electric field $\vec{B} = \vec{\mathcal{B}}/c$ so that we have

$$\nabla \times \vec{\mathcal{B}} - \epsilon_r \frac{\partial \vec{E}}{\partial \tau} = 0 \tag{1.90}$$

$$\nabla \times \vec{E} + \frac{\partial \vec{\mathcal{B}}}{\partial \tau} = 0, \tag{1.91}$$

where we used $c^2 = 1/\epsilon_0 \mu_0$. We use a comparison with the block (with mass m) attached to a spring with constant k. The block dynamic is described by the coordinate x and momentum p. We find

$$p - m\frac{dx}{dt} = 0, \tag{1.92}$$

$$kx + \frac{dp}{dt} = 0; \tag{1.93}$$

here we adopt a somewhat 'unusual' notation to emphasize the similarity between the electromagnetic and mechanical system. That is, Faraday's law corresponds to the definition of momentum in terms of the velocity, and the magnetic field produced by the tie-dependent displacement current corresponds to Newton's second law. We conclude that the dielectric constant corresponds to inertia; this suggests that the presence of dielectric matter slows down the propagation of electromagnetic fields.

For a mathematical derivation, we take the curl of Eq. (1.91) and find $-\nabla^2\vec{E} + (\partial/\partial t)\nabla \times \vec{B} = 0$ so that Eq. (1.90) yields $-\nabla^2\vec{E} + (\partial/\partial t)[\mu_0 \partial\vec{D}/\partial t] = -\nabla^2\vec{E} + \epsilon_r c^{-2}(\partial^2\vec{E}/\partial t^2) = 0$, where we introduced the speed of light in vacuum $c = 1/\sqrt{\epsilon_0\mu_0}$. Alternatively, we take the curl of Eq. (1.90) and find $-\nabla^2 H - (\partial/\partial t)[\nabla \times \vec{D}] = 0$ so that using Eq. (1.91) $-\nabla^2 H + (\partial/\partial t)[\epsilon_r\epsilon_0 \partial B/\partial t] = \mu_0^{-1}\left[-\nabla^2 B - \epsilon_r c^{-2}(\partial^2\vec{B}/\partial t^2)\right] = 0$. We introduce the index of refraction

$$n = \sqrt{\epsilon_{\rm r}} > 1, \tag{1.94}$$

and the speed of light in matter

$$v = \frac{c}{n} < c. \tag{1.95}$$

We arrive at the wave equation for the electric and magnetic field,

$$\nabla^2 \left\{ \begin{array}{c} \vec{E} \\ \vec{B} \end{array} \right\} - \frac{1}{v^2}\frac{\partial^2}{\partial t^2} \left\{ \begin{array}{c} \vec{E} \\ \vec{B} \end{array} \right\} = 0. \tag{1.96}$$

The wave equation has solutions traveling along the $+\hat{\mathsf{k}}$ and $-\hat{\mathsf{k}}$- directions, respectively (we use sans-serif k for wave vectors to distinguish from spring constant k, see Sect.1.4). We write $\vec{\mathsf{k}} = \mathsf{k}\hat{\mathsf{k}}$ and $v\,\mathsf{k} = v2\pi/\lambda = 2\pi f = \omega$ and find $\vec{E}(z,t) = \vec{E}_{\mp}(\vec{\mathsf{k}}\cdot\vec{r} \mp \omega t)$ and $\vec{B}(z,t) = \vec{B}_{\mp}(\vec{\mathsf{k}}\cdot\vec{r} \mp \omega t)$. A 'special' (or particular) solution of the wave equation are plane waves

$$\left\{ \begin{array}{c} \vec{E}(z,t) \\ \vec{B}(z,t) \end{array} \right\} = \left\{ \begin{array}{c} \vec{E}_{0,\mp} e^{i(\vec{\mathsf{k}}\cdot\vec{r}\mp\omega t)} \\ \vec{B}_{0,\mp} e^{i(\vec{\mathsf{k}}\cdot\vec{r}\mp\omega k t)} \end{array} \right\}. \tag{1.97}$$

From Eqs. (1.90) and (1.91), we have

$$\vec{\mathsf{k}}\cdot\vec{E} = 0, \quad \vec{\mathsf{k}}\cdot\vec{B} = 0 \tag{1.98}$$

so that the electric and magnetic fields are perpendicular to the direction of propagation: electromagnetic waves are *transverse* waves. The curl equations yield

$$\vec{B} = \sqrt{\mu\epsilon}\vec{\mathsf{k}} \times \vec{E}. \tag{1.99}$$

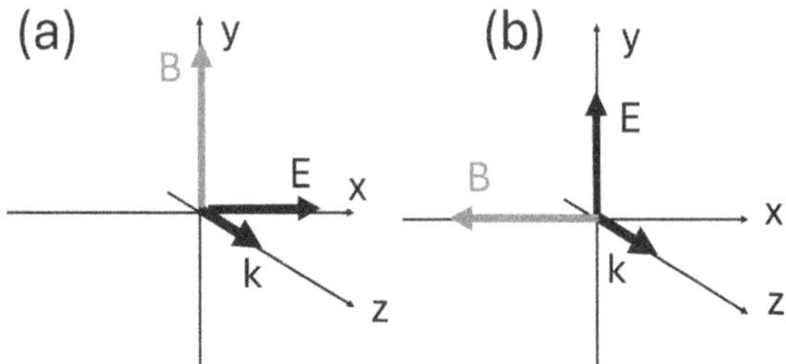

FIGURE 1.8
Linearly polarized electromagnetic waves propagating along the $+z$-axis. The two possible polarizations correspond to (a) the electric (black) and magnetic (gray) fields pointing along the $+x$- and $+y$-axes, respectively, and (b) the electric and magnetic fields pointing along the $+y$- and $-x$-axes, respectively.

The mutual orthogonality of the electric field $\vec{E}$, the magnetic field $\vec{B}$, and the wave vector $\vec{\mathsf{k}}$ are shown in Fig. 1.8. Under spatial inversion, we have $\vec{\mathsf{k}} \to \vec{\mathsf{k}}' = -\vec{\mathsf{k}}$ and $\vec{E} \to \vec{E}' = -\vec{E}$ so that $\vec{\mathsf{k}} \times \vec{E} \to \vec{\mathsf{k}}' \times \vec{E}' = \vec{\mathsf{k}} \times \vec{E}$ which is consistent with the axial character of the magnetic field $\vec{B}$, cf. Fig. 1.9.

If we choose a coordinate system such the wave vector is aligned with the z-axis $\vec{\mathsf{k}} = \mathsf{k}\hat{z}$, the electric and magnetic field are in the (x, y)-plane: (i) $\vec{E} = E_0\hat{x}$ and $\vec{B} = \sqrt{\mu\epsilon}E_0\hat{\jmath}$ and (ii) $\vec{E} = E_0\hat{y}$ and $\vec{B} = -\sqrt{\mu\epsilon}E_0\hat{x}$.

In spectroscopy, the customary unit for frequency is inverse centimeter $[f] = \text{cm}^{-1}$; that is, it corresponds to a frequency of an electromagnetic wave with wavelength $\lambda = 1\,\text{cm}$ so that $f = (3.0 \times 10^8\,\text{m/s})/(1 \times 10^{-2}\text{m})$ or

$$f = 1.0\,\text{cm}^{-1} = 3.0 \times 10^{10}\text{Hz}, \tag{1.100}$$

which corresponds to $f = 7.5 \times 10^{-4}\,\nu$.

Polarization: The electric field (and thus the corresponding magnetic field) is directed along a fixed direction in the plane perpendicular to the direction of propagation (taken as the z-axis), cf. Fig. 1.10. This allows for two perpendicular polarizations which we take as x- and y-axis: $\vec{E} = (E_x\hat{x} + E_y\hat{y})\exp(i[\mathsf{k}z - \omega t])$. In general, the amplitudes are complex-valued and characterized by an amplitude and phase. One phase can be chosen arbitrarily: we set $\phi_x = 0$ so that the x-component is real $E^*_{x,0} = E_{x,0}$. We write $\phi_y = \phi$ and

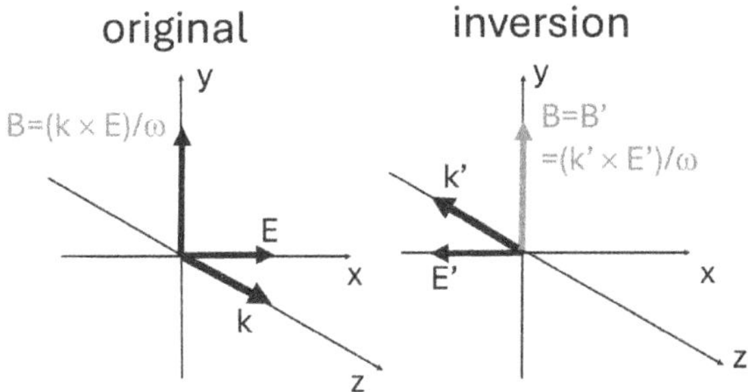

FIGURE 1.9
In the original version, the wave propagates long the $+z$-axis and the electric field $\vec{E}$ (black) points along the $+x$-axis. In the inverted version, the wave propagates along the $-z$-axis and the electric field points along the $-x$-axis. Because the magnetic field $\vec{B}$ (gray) is unchanged, the original version corresponds to *right*-circularly polarized light, while the inverted version corresponds to *left*-circularly polarized light.

TABLE 1.1
Spectrum of the electromagnetic waves in vacuum frequency f in unit $[f] = \text{cm}^{-1}$ and $[f] = \text{cm}^{-1}$.

Spectrum	f [s^{-1}]	f [cm^{-1}]
Microwave	$3 \times 10^9 - 3 \times 10^{12}$	$0.1 - 100$
Infrared	$3 \times 10^{12} - 4 \times 10^{14}$	$100 - 1.4 \times 10^4$
Visible	$4 \times 10^{14} - 7.5 \times 10^{14}$	$1.4 \times 10^4 - 2.5 \times 10^4$
Ultraviolet	$7.5 \times 10^{14} - 3 \times 10^{17}$	$2.5 \times 10^4 - 1 \times 10^7$

$E_y = E_{y,0} e^{i\phi}$,

$$\vec{E} = (E_{x,0}\hat{x} + E_{y,0}e^{i\phi}\hat{y})e^{i(kz-\omega t)}. \tag{1.101}$$

We discuss two cases characterized by the phases $\phi = 0$ and $\phi = \pm\pi/2$. (1) Light is linearly polarized with magnitude of the electric field is $E_0 = \sqrt{E_{x,0}^2 + E_{y,0}^2}$ at the angle $\tan\theta = E_{y,0}/E_{x,0}$ with respect to the x-axis.
(2) For $\phi = \pi/2$, we have $e^{\pm i\pi/2} = \pm i$; we recall that a multiplication by the imaginary unit i is equivalent with a

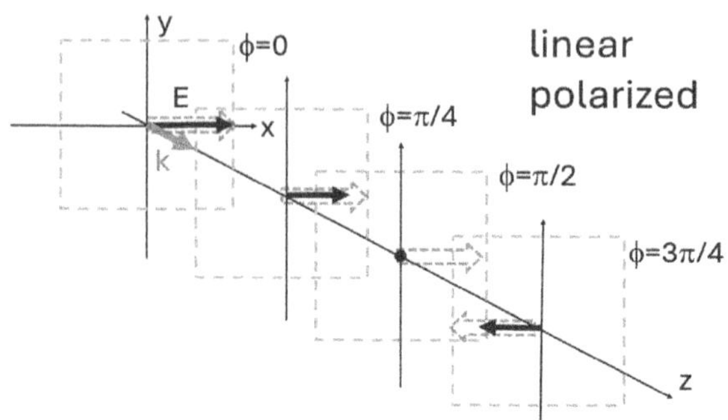

FIGURE 1.10
Linear polarized light; the light travels along the $+z$-axis $\vec{k} = k\hat{z}$ and the electric field $\vec{E}$ is directed along the x-axis, $\vec{E} = E_0 \cos\phi \cdot \hat{x}$ for the phase $\phi = kz - \omega t$ shown for $\phi = 0, \pi/4, \pi/2$, and $3\pi/4$.

rotation by 90° in either counterclockwise (+) or clockwise direction. We write $E_x = E_0$ and $E_y = \pm iE_0$ so that $\vec{E}(z,t) = E_0(\hat{x} \pm i\hat{y}) \exp[i(\mathsf{k}z - \omega t)]$; that is, electric field varies between the x- and y-axes: $\vec{E} = E_0[\cos(\mathsf{k}z - \omega t)\,\hat{x} + \cos(\mathsf{k}z - \omega t + \pi/2)\,\hat{y}]$, or

$$\vec{E}(z,t) = E_0\left[\cos(\mathsf{k}z - \omega t)\,\hat{x} \mp \sin(\mathsf{k}z - \omega t)\,\hat{y}\right]. \tag{1.102}$$

We examine the electric field at a constant point $z = z_0$ and 'look toward the oncoming wave.' The case $\hat{x} + i\hat{y}$ ($\hat{x} - i\hat{y}$) corresponds to a *counterclockwise* (*clockwise*) rotation with frequency ω: this is referred to as a left- (right-) circularly polarized wave in the context of optics (which we follow here). However, in the context of a description of electromagnetic waves in terms of photon, left- (right-) circularly polarized photons have positive (negative) helicity. Right- and left-circularly polarized light is sketched in Fig. 1.11. The rotation of the electric field at a fixed location is shown in Fig. 1.12.

Eq. (1.102) states that circularly polarized light is the superposition of linearly polarized light. The reverse is also the case, cf. Fig. 1.13. At time t, the electric fields associated with right- and left-polarized light have both x and y components

$$E_{r,x}(t) = E_0 \cos(\omega_r t) \qquad E_{r,y}(t) = -E_0 \sin(\omega_r t), \tag{1.103}$$
$$E_{l,x}(t) = E_0 \cos(\omega_l t) \qquad E_{l,y}(t) = E_0 \sin(\omega_l t), \tag{1.104}$$

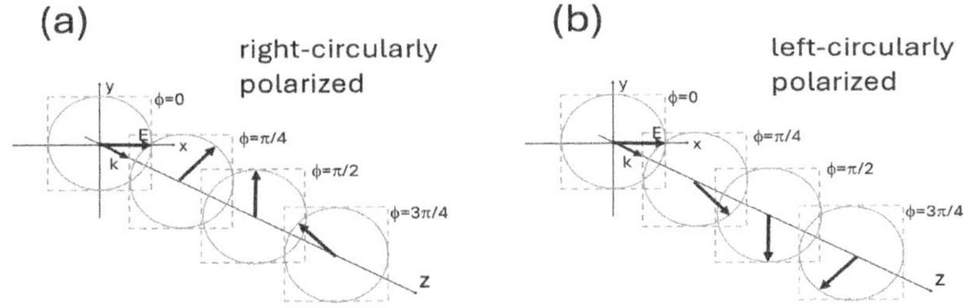

FIGURE 1.11
Circularly-polarized light at some fixed time t. The light travels along the $+z$-axis $\vec{k} = k\hat{z}$ (out-of-the-page). At $z = 0$ the electric field is directed along the x-axis, $\vec{E}(z = 0) = E_0\hat{x}$. The electric field is shown for the phase $\phi = kz = 0, \pi/4, \pi/2,$ and $3\pi/4$. (**a**) Right-circularly polarized light (negative helicity) and (**b**) left-circularly polarized light (positive helicity).

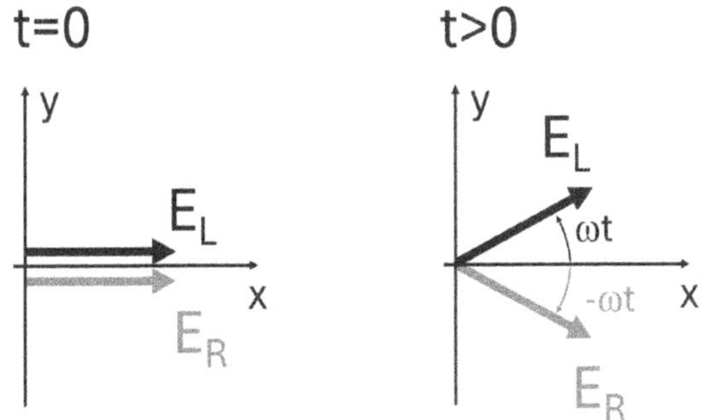

FIGURE 1.12
Circularly-polarized light at some fixed position z. The light travels along the $+z$-axis $\vec{k} = k\hat{z}$ (out-of-the-page). At $t = 0$ the electric field is directed along the x-axis, $\vec{E}(t = 0) = E_0\hat{x}$. For $t > 0$, the electric field $\vec{E}(t)$ rotates *clockwise* such that the phase is negative ($\theta = -\omega t < 0$) for right-circularly polarized light (gray), while the electric field rotates *counterclockwise* such that the phase is positive ($\theta = -\omega t > 0$) for left-circularly polarized light (black).

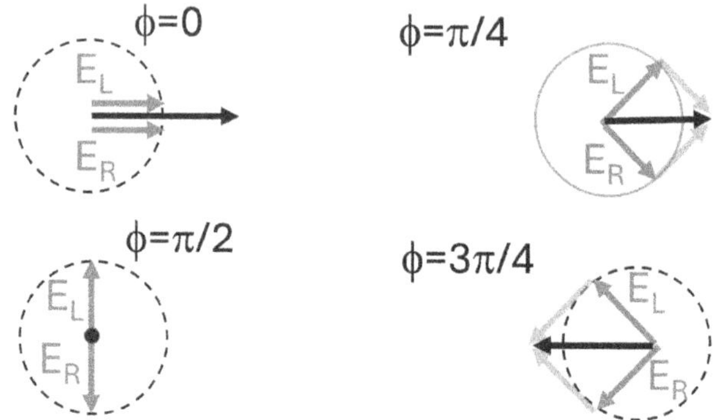

FIGURE 1.13
Linearly polarized light (black) as a superposition of left- and right-circularly (both gray) polarized light for $\phi = 0, \pi/4, \pi/2$, and $3\pi/4$.

where we choose the time $t = 0$ such that an arbitrary phase is zero. We find for the superposition $\vec{E} = \vec{E}_r + \vec{E}_l$ with

$$E_x(t) = E_{r,x}(t) + E_{l,x}(t) = 2E_0 \cos(\omega t) \tag{1.105}$$
$$E_y(t) = E_{r,y}(t) + E_{l,y}(t) = 0, \tag{1.106}$$

that is, we recover linearly polarized light along the y-axis.

We write the electric field in the (x, y)-plane in complex notion with

$$\mathrm{Re}E = E_x \quad \mathrm{Im}E = E_y. \tag{1.107}$$

We describe the time-dependence in terms of the complex exponential factor $\exp(i\omega t)$. We then have for circularly polarized light

$$E_r(t) = \frac{1}{\sqrt{2}} \begin{pmatrix} 1 \\ -i \end{pmatrix} e^{-i\omega t} \quad \text{right-circular,} \tag{1.108}$$
$$E_l(t) = \frac{1}{\sqrt{2}} \begin{pmatrix} 1 \\ i \end{pmatrix} e^{-i\omega t} \quad \text{left-circular.} \tag{1.109}$$

We describe the electric and magnetic fields, $\vec{E}$ and $\vec{B}$, and their derivates, $\partial\vec{E}/\partial t$ and $\partial\vec{B}/\partial t$, for right- and left-circularly polarized light at a fixed coordinate $z = z_0$. We assume that at time t, the electric field is directed along the $+x$ axis, $\vec{E}(t) = E_0\hat{x}$. The magnetic field is directed along the $+y$-axis, $\vec{B} = B_0\hat{y}$. Since

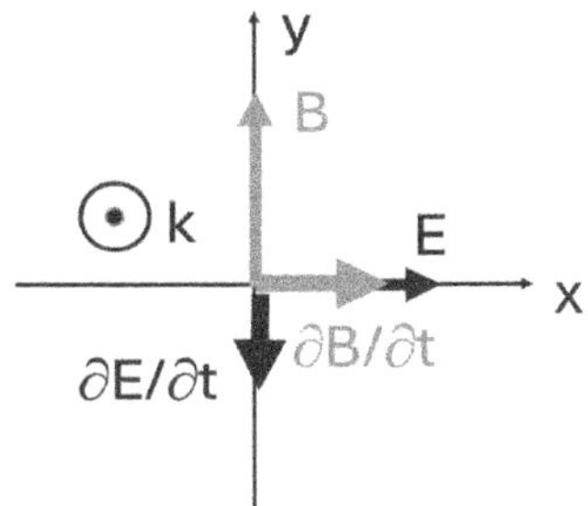

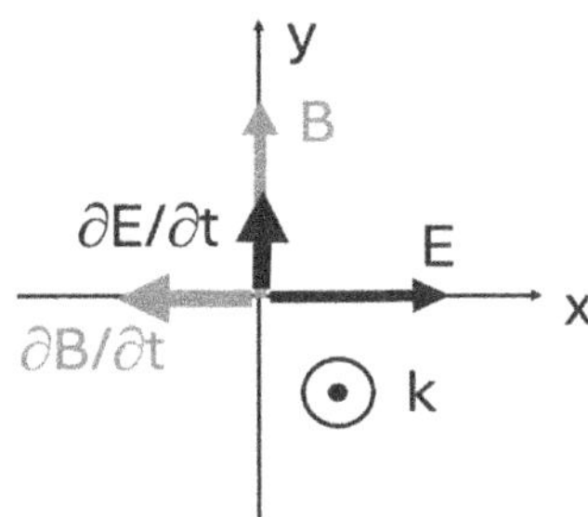

FIGURE 1.14
The electric (black) and magnetic (gray) field $\vec{E}$ and $\vec{B}$ and their time derivatives $\partial\vec{E}/\partial t$ (black) and $\partial\vec{B}/\partial t$ (gry) for right- and left-circularly polarized light. The direction of propagation (the vector $\vec{\mathsf{k}}$) is directed out-of-the-page.

the field vectors sweep *clockwise* (*counterclockwise*) for right- and left-circularly polarized light, the electric field $\vec{E}(t+\Delta t)$ is directed in the fourth (first) quadrant and the magnetic field $\vec{B}(t+\Delta t)$ is directed in the first (second) quadrant. It follows that for right-circularly polarized light $\partial\vec{E}/\partial t$ is antiparallel to $\vec{B}$ and $\partial\vec{B}/\partial t$ is parallel to $\vec{E}$; for left-circularly polarized light $\partial\vec{E}/\partial t$ is antiparallel to $\vec{B}$ and $\partial\vec{B}/\partial t$ is parallel to $\vec{E}$, cf. Fig. 1.14.

Symmetry considerations often provide valuable insight into properties of physical systems. For electromagnetic waves, the relevant symmetry is spatial inversion $\vec{r} \to -\vec{r}$. The wavevector and electric field are polar vectors that change signs: $\vec{\mathsf{k}} \to -\vec{\mathsf{k}}$ and $\vec{E} \to -\vec{E}$. The magnetic field is an axial vector and this does not change sign $\vec{B} \to \vec{B}$. Taking the time derivative does not change the symmetry under inversion so $\partial\vec{E}/\partial t \to -\partial\vec{E}/\partial t$ and $\partial\vec{B}/\partial t \to \partial\vec{B}/\partial t$. Fig. 1.15 shows the transformation of right- and left-circularly polarized light: we see by inspection that *right*-circularly polarized light becomes *left*-circularly polarized light and vice versa. This shows that in systems with inversion symmetry, there is no difference between the right- and left-circularly polarized electromagnetic waves. Or put differently, any difference

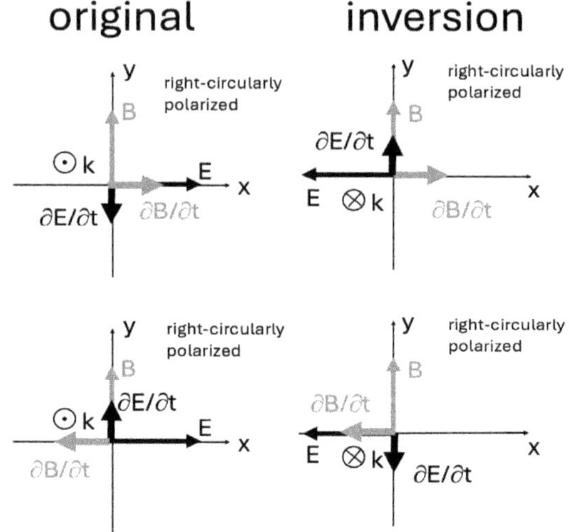

FIGURE 1.15
Right- and left-circularly polarized light in the original (x, y, z) and inverted $(x \rightarrow -x, y \rightarrow -y, z \rightarrow -z)$ systems.

between right- and left-circular polarized light is only possible in systems without inversion symmetry.

Light Absorption: Light intensity is generally diminished as it travels through dielectric matter. The Beer-Lambert law states that the absorbance is proportional to the concentration of the absorbing substance. In a macroscopic description, light absorption can be described in terms of complex-valued wave vector

$$\mathsf{k} \longrightarrow \mathsf{k}' + i\,\mathsf{k}'' \tag{1.110}$$

with $\mathsf{k}' = \mathrm{Re}(\mathsf{k})$ and $\mathsf{k}'' = \mathrm{Im}(\mathsf{k})$. That is, a spatially varying electric field becomes exponentially decaying

$$e^{ikx} \longrightarrow e^{i\mathsf{k}'x}e^{-\mathsf{k}''x}. \tag{1.111}$$

Here, the imaginary part k'' is determined by the interaction of light with matter; in particular, light absorption involves a jump of an electron from its ground state to an excited state. While we are primarily interest in absorption due to molecules, we gain some insight by first discussing absorption due to atoms.

The absorption of light by atoms is discussed in introductory physics and chemistry texts mostly as it relates to the explanation of line spectra superimposed on the continuous spectrum. The description of a multi-electron atom can typically simplified by only considering the outer most electron (so-called valence electron); the remaining electrons play either a minor or no role at all. In our discussion, 'electron' refers to the valence electron. The explanation of line spectra in terms of discrete electron 'orbits' played a central role in the development of *Quantum Mechanics* in the early 20th century. A proper treatment involves finding solutions of a appropriate time-independent Schrödinger equation. For a qualitative discussion, a semiclassical description in terms of Bohr's quantization condition is often adequate: the quantization can be expressed as the condition that de Broglie wavelength of the electron forms a standing wave on its orbit. A bigger radius of the orbit corresponds to an electron that is more loosely bound electron, that is, to a higher energy eigenvalue of the electron. There is a one-to-one correspondence between a discrete absorption/emission line and a jump of the electron from one such level to another. The jumps occur on time scales much shorter than any other relevant time (e.g., the period in description of the electron orbit in a classical description): the jumps are considered as instantaneous.

1.5 Molecular Systems

Units: Both gram and kilogram are appropriate units of mass for macroscopic units. Atoms and molecules are microscopic objects; it is useful and customary to use 'conventional' units; however, the choice is not unique, and varies between scientific disciplines. It is quite standard in chemistry to convert to *molar* quantities which are obtained by a quantities for a single atom or molecule by Avogadro's number $N_A = 6.022 \times 10^{23}\,\mathrm{mol}^{-1}$. The atomic mass unit (amu), or dalton (Da) is defined via the mass of a single ^{12}C atom $m_{\mathrm{C}} = 12.0\,\mathrm{amu}$. One mole is a unit so that 1 gram cannot be compared to a 1 gram per mol in a strict sense. If we continue to be a bit sloppy, similar to units in electromagnetism, we write

$$\mathcal{M} = 1\,\mathrm{amu} = 1.6605 \times 10^{-27}\,\mathrm{kg}. \tag{1.112}$$

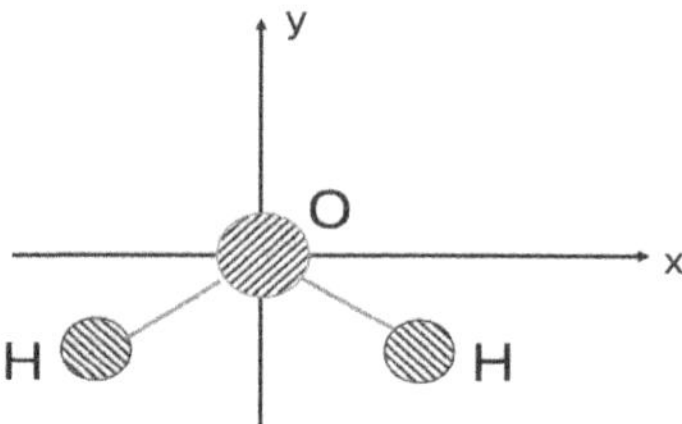

FIGURE 1.16
Geometry of the water molecule (H_2O). The x- and z-axes are in the plane of the molecule; the y axis is perpendicular to the plane of the molecule.

The typical size of atoms and bond lengths of molecules is of the order of $1\,\text{Å} = 10^{-10}\,\text{m}$; for example, the Bohr radius characterizing the electron orbits is $a_0 = 0.529\,\text{Å}$ and the OH-bond of a water molecule (H_2O) is $d = 0.96\,\text{Å}$. The geometry of a water molecule is shown in Fig. 1.16. We have the length scale

$$\mathcal{L} = 1\,\text{Å} = 0.1\,\text{nm}. \tag{1.113}$$

In liquids and gases, molecules tumbles more-or-less freely. Molecules can be described by rigid rotors; the angular momentum is quantized $|\vec{L}| = (h/2\pi)\sqrt{j(j+1)}$ with $j = 1, 2, ...,$ the moment of inertia is of the order $I = \mathcal{M}\mathcal{L}^2 \sim 1.67 \times 10^{-27}\,\text{kg} \cdot (10^{-10}\,\text{m})^2 \sim 2 \times 10^{-47}\,\text{kg}\,\text{m}^2$. For example, for the H_2O molecule, the moment of inertia about the three (perpendicular) axes are $I_{xx} = 1.00 \times 10^{-47}\text{kg} \cdot \text{m}^2$, $I_{yy} = 2.92 \times 10^{-47}\text{kg} \cdot \text{m}^2$, and $I_{zz} = 1.91 \times 10^{-47}\text{kg} \cdot \text{m}^2$.

The electric charge of electrons is the elementary charge

$$e = 1.602 \times 10^{-19}\,\text{C} = 4.8 \times 10^{-10}\,\text{statC}. \tag{1.114}$$

In the cgs system, it is quite usual to define the so-called *electrostatic unit of charge*

$$1\,\text{esu} = 10^{-10}\,\text{statC} \tag{1.115}$$

so that one elementary charge is $e = 4.8\,\text{esu}$. While molecules have. generally no net charge, they have non-trivial charge distribution at lengths of the order $\mathcal{L}$. The relevant charges relevant are multiples or fractions of the elementary charge, For molecular systems, the bonding of elements with different electronegativity can be expressed in terms of a *partial charge* $q = \pm\delta e$ that describe the chemical bonding as a continuum between ionic ($\delta = 1$) and covalent ($\delta = 0$). The electric dipole moment is defined by a positive $+q$ and a negative charge $-q$ separated by the distance d, $p = qd$; the vector is directed from the *negative* to the *positive* charge. The length scale $\mathcal{L}$ and the elementary charge e define the dipole moment

$$p = e\mathcal{L} = 1.602 \times 10^{-29}\,\mathrm{C\,m}. \tag{1.116}$$

In many references, especially in the 'older' literature, the unit of *debye* (D), based on the statC, or esu is used

$$1\,\mathrm{D} = 1\mathrm{esu} \cdot \mathcal{L} = 3.335 \times 10^{-30}\,\mathrm{Cm}. \tag{1.117}$$

We then have $e\mathcal{L} = 4.823\,\mathrm{D}$. For the water molecule H_2O, for example, the partial charges are $\delta_{\mathrm{H}} = 0.335$ and $\delta_{\mathrm{O}} = -2\delta_H = -0.67$, respectively. The OH bond length is $d = 9.42 \times 10^{-11}\,\mathrm{m} = 0.942\,\mathcal{L}$; the angle between the two bonds is 104.5°. The dipole moment follows $p_{\mathrm{water}} = 2 \cdot 0.942 \cdot 0.335 \cdot \cos(104.5°/2) \cdot e\,\mathcal{L}$ so that $p_{\mathrm{water}} = 6.08 \times 10^{-30}\,\mathrm{C\,m} = 1.84\,\mathrm{D}$.

An energy scale for molecular systems is not uniquely defined; however, different definitions yield roughly the same value so that the exact definition is irrelevant. Here, we have a definition based on the commonly used energy for 'chemical' energies (i.e., *free* energy)

$$\mathcal{U} = 1\,\frac{\mathrm{kJ}}{\mathrm{mol}} = 1.67 \times 10^{-21}\,\mathrm{J}. \tag{1.118}$$

The thermal energy k_BT (where $k_B = 1.381 \times 10^{-23}\,\mathrm{J/K}$ is the Boltzmann constant) corresponding to room temperature ($T = 300\,\mathrm{K}$) follows $U = 4.14 \times 10^{-21}\,\mathrm{J} = 2.5\mathcal{U}$. In the physics literature, an *electronvolt* is used $1\,\mathrm{eV} = 1.609 \times 10^{-19}\,\mathrm{J} = 96.6\mathcal{U}$. Molecules 'tumble' around axes of rotation: the angular momentum is quantized such that $|\vec{L}| \simeq l\hbar$ (where $\hbar = h/2\pi = 1.0 \times 10^{-34}\mathrm{J\,s}$ is the reduced Planck's constant). For $l = 1$, the kinetic energy follows $\mathrm{KE_{rot}} \simeq (1.0 \times 10^{-34}\mathrm{J\,s})^2/(2 \cdot 2.0 \times 10^{-47}\,\mathrm{kg\,m^2}) = 5.0 \times 10^{-22}\,\mathrm{J} = 0.3\mathcal{U}$.

The interaction of two molecules with permanent dipole moments separated by the distance $d \simeq 2\mathcal{L}$ follows $U \simeq (4\pi\epsilon_0)^{-1}(1\,\mathrm{D})^2/\mathcal{L}^3 = 10^{10}(\mathrm{Nm^2/C^2}) \cdot (10^{-30}\,\mathrm{Cm})^2/(2 \cdot 10^{-10}\,\mathrm{m})^3 = 10^{-21}\,\mathrm{J} \simeq 1\mathcal{U}$. These estimates are consistent with the known properties of water H_2O). The dipole-dipole interaction leads to attractive forces between neighboring molecule (hydrogen bond), which, in turns, can yield large scale structures (e.g., snowflakes). At room temperatures, the bonds break and the crystal (solid) becomes a liquid and the molecules tumble freely. Water is somewhat unique and for most dielectric materials, the dipole-dipole interaction is too weak to cause long-range (configurational) correlations between molecules so that their dielectric properties reflect the average properties of individual molecules.

For the molecular time scale, we use energy $=$ mass $\cdot$ length^2/time^2 and find

$$\sqrt{\frac{\mathcal{M}\cdot\mathcal{L}^2}{\mathcal{U}}} = \sqrt{\frac{(1.67\times 10^{-27}\,\text{kg})\cdot(10^{-10}\,\text{m})^2}{(1.66\times 10^{-21}\,\text{J})}} = 1.0\times 10^{-13}\,\text{s}.$$

Find a similar estimate from Planck's constant ($h = 6.64\times 10^{-34}$ Js) and the energy scale,

$$\frac{h}{\mathcal{U}} = \frac{(6.64\times 10^{-34}\,\text{Js})}{(1.66\times 10^{-21}\,\text{J})} = 4.0\times 10^{-13}\,\text{s}.$$

We choose for time scale

$$\mathcal{T} = 1.0\times 10^{-13}\,\text{s} = 0.1\,\text{ps}, \tag{1.119}$$

and arrive at the characteristic frequency $\nu = 1/\mathcal{T} = 1.\times 10^{13}\,\text{s}^{-1}$. This corresponds to an electromagnetic wave with wavelength $\lambda = (3.0\times 10^8\,\text{m/s})/(1.0\times 10^{13}\,\text{s}^{-1}) = 30\,\mu\text{m}$; or in terms of wave numbers $\nu = 333\,\text{cm}^{-1}$,

The characteristic energy $\mathcal{U}$ and length scale $\mathcal{L}$, yields the characteristic force

$$\mathcal{F} = \frac{\mathcal{U}}{\mathcal{L}} = 1.67\times 10^{-11}\,\text{N} = 1.67\times 10^{-6}\,\text{dyn}. \tag{1.120}$$

The characteristic spring (or elastic) constant follows from $\mathcal{U} = \frac{1}{2}\mathcal{K}\mathcal{L}^2$ so that $2\mathcal{U}/\mathcal{L}^2 = (2{\cdot}1.67{\times}10^{-21}\,\text{J})/(10^{-10}\,\text{m})^2 = 0.334\,\text{N/m}$. Alternatively, the characteristic spring constant can be defined by the mass $\mathcal{M}$ and the characteristic angular frequency $2\pi\,\nu$: $\mathcal{K} = \mathcal{M}(2\pi\nu)^2 = 1.67{\times}10^{-27}\,\text{kg}{\cdot}(6.28{\times}10^{13}\,\text{rad/s})^2 = 6.6\,\text{N/m}$. We see that the two estimates are consistent with each other; we choose the characteristic scale

$$\mathcal{K} = 1\frac{\text{N}}{\text{m}} = 1,000\,\frac{\text{dyn}}{\text{cm}}. \tag{1.121}$$

Oscillations in molecules: The (geometric) structure of molecules are described by chemical bonds; their elastic properties can be described by springs with constant k. For the diatomic molecule carbonmonoxide (CO), we have $2 \times 3 = 6$ degrees of freedom; three degrees of freedom correspond to the center of mass (CoM) and two degrees of freedom correspond to rotations around axes (y- and z-axis) perpendicular to axis of the molecule (x-axis). The rotation around the x-axis is 'frozen out.' The motion of atoms relative to the (fixed) CoM is described by a fictitious particle with 'reduced' mass $\mu^{-1} = m_{\mathrm{C}}^{-1} + m_{\mathrm{O}}^{-1}$. The (angular) frequency follows $\omega = \sqrt{k/\mu}$. Carbonmonoxide is *IR active* with a frequency $f = 2,143\,\mathrm{cm}^{-1} = 6.43 \times 10^{13}\,\mathrm{Hz}$. The reduced mass is $\mu = 6.86\,\mathrm{amu} = 1.15 \times 10^{-26}\,\mathrm{kg}$. The spring (elastic) constant follows $k_{\mathrm{CO}} = (2\pi\mathrm{rad} \cdot 6.43 \times 10^{13}\,\mathrm{s})^2 \cdot 1.15 \times 10^{-26}\,\mathrm{kg} = 1,877\,\mathrm{N/m} = 1.877 \times 10^6\,\mathrm{dyn/cm}$ or in terms of the characteristic elastic constant $\mathcal{K}$, cf. Eq. (1.121), $k_{\mathrm{CO}} = 58.3\,\mathcal{K}$.

For larger molecules, the motion of the atoms (ions) is described by more than one vibrational mode. In particular, several bonds vibrate in 'synchrony' with a pattern, governed by the symmetry of the molecule. We illustrate the connection between modes and symmetry using the simple (but widely familiar) example of carbondioxide (CO_2), cf. Fig. 1.17. The molecular axis is taken to correspond to the x-coordinate axis so that the equilibrium coordinates are $x_1^0 = -a$, $x_2^0 = 0$, and $x_3^0 = a$ with $a = 0.1163\,\mathrm{nm}$. We write $x_{1,3} = x_{1,3}^{(0)} + x'_{1,3}$ and drop the prime.

We exclude motion of the CoM and rotations about the y and z-axes so that the remaining $(3 \times 3 - 3 - 2 = 4)$ degrees of the freedom describe vibrations of the molecule. For small displacements of the springs from their equilibrium lengths, the elastic potential energies are the sum of contributions along the three coordinate axes $(k/2)\left[(x_{1,3} - x_2)^2 + (y_{1,3} - y_2)^2 + (z_{1,3} - z_2)^2\right]$. It follows that vibrations along two perpendicular coordinate axes are independent of each other; two degenerate modes are associated with vibrations along the y- and z-axis and two modes are associates are associated with vibrations along the x-axis.

We examine the motion along the x-axis. The condition of a fixed CoM reads $m_{\mathrm{O}}x_1 + m_{\mathrm{C}}x_2 + m_{\mathrm{O}}x_3 = 0$, so that the motion of the central carbon atom is already determined by the motion of the two oxygen atoms $x_2 = -(m_{\mathrm{O}}/m_{\mathrm{C}})(x_1 + x_3)$; this introduces a coupling of the coordinates or velocities of the oxygen atoms $\delta U \sim x_1 x_3$ and $\delta U' \sim \dot{x}_1 \dot{x}_3$, respectively. We introduce the sum and difference coordinates

$$q_a = x_3 + x_1 \qquad \text{and} \qquad q_s = x_3 - x_1. \tag{1.122}$$

For $q_a \neq 0$ and $q_s = 0$, both oxygen atoms move in the same direction and the carbon atom moves in opposite direction; that is, one bond is stretched while the other one is squeezed, corresponding to the antisymmetric mode. For $q_a = 0$ and $q_s \neq 0$, the oxygen atoms move in the opposite directions and the carbon atom remains at rest; that is, the two bonds are either stretched or squeezed (depending on the phase) - corresponding to the antisymmetric mode.

The energy can be written as the sum of the energy of the symmetric and antisymmetric mode $U(\{x_i, \dot{x}_i\}_{i=1,2,3}) \to U(q_a, \dot{q}_a) + U(q_s, \dot{q}_s)$ with $U(q_{a,s}, \dot{q}_a) = m_{a,s}\dot{q}_{a,s}^2/2 + k_{a,s}q_{a,s}^2/2$, where we introduced masses $m_a = (m_\text{O}M/2m_\text{C})$ and $m_s = m_\text{O}/2$ and spring constants $k_a = k(M/m_\text{C})^2$ and $k_s = k/2$ (and $M = 2m_\text{O} + m_\text{C}$ is the total mass). We find the values $\omega_s = 1340\,\text{cm}^{-1}$ and $\omega_a = 2349\,\text{cm}^{-1}$.

For the two degenerate modes along the y- and z-axis, we write $y_1 = y_3 = y$ and $y_2 = -(2\,m_\text{O}/m_\text{C})y$. Bending involves a rotation of the two C-O bonds in opposite directions (clockwise and counterclockwise) at angles $\theta \simeq \pm(M/m_\text{C}) \cdot (a/y)$ around their respective centers. The distances of the atoms from the center of rotation are $l_\text{O} = (m_\text{C}/M)a$, and $l_\text{C} = (2m_\text{O}/M)a$. We then have for small displacements, $y = l_\text{O}\theta$ and $y_2 = l_\text{C}\theta$. The energy for the motion along the y-axis can then be written $U(\{y_i, \dot{y}_i\}_{i=1,2,3}) \to U(\theta, \dot{\theta})$ with $U(\theta, \dot{\theta}) = Ml_\text{O}l_\text{C}\dot{\theta}^2/2 + kl_\text{O}a\theta^2/4$. The (angular) frequency of bending follows $\omega_b = (1/2)\sqrt{k/m_\text{O}} = \omega_s/2$ so that $\omega_b = 667\,\text{cm}^{-1}$.

Independence of modes: A coupling of two (vibrational) modes implies that there is a transfer from one mode to the other. Applied to the motion of the CO_2 molecule along the x-axis, this implies that the system can start-off vibrating with the stretching mode excited and the antisymmetric stretch suppressed to evolve such that it vibrates with the antisymmetric stretch excited and the symmetric stretch suppressed. That is, the system starts off with the two bonds vibrating in synchronous fashion (i.e., both bonds compressed or stretched) and ending with the two bonds vibrating in asynchronous fashion (i.e., the two bonds compressed/stretched or stretched/compressed). We conclude that the coupling between symmetric and antisymmetric stretch must not have a distinct symmetric; for vibrations along the x-axis, this implies that the coupling $c(x)$ must have both an even and odd part $c(x) = c_+(x) + c_-(x)$ with $c_+(-x) = c_+(-x)$ and $c_-(-x) = -c_-(x)$.

A coupling of the bending mode (along the y-axis, say) and either the symmetric or antisymmetric stretch would require not only

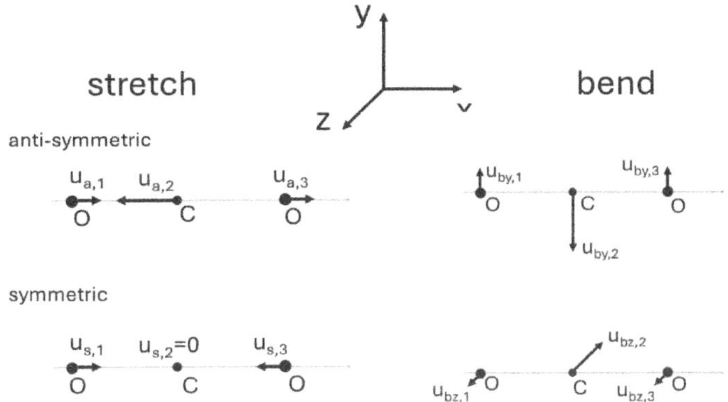

FIGURE 1.17
The normal modes (independent vibrations) for the linear molecule CO_2: antisymmetric and symmetric stretch in bend (along two orthogonal directions).

a transfer of energy between modes but also the action of torque on the system as motion along the y-axis when the bending mode is excited (and the symmetric/antisymmetric stretch is suppressed) to a motion along the x-axis when the symmetric/antisymmetric stretch is excited (and the bending mode is suppressed). For the CO_2 molecule, we model the bonds as elastic springs with stiffness along the spring but no *torsional* stiffness. In fact, the forces associated with the symmetric/antisymmetric stretch are directed along the x-axis and thus cannot do work to displacements associated with the bending mode along the perpendicular y-axis. In this case, work is done by torques τ, $W = \tau \cdot \Delta\theta$, where $\Delta\theta$ is the angular displacement.

For the torques around the $+z$-axis (out-of-the-page) and thus rotations in the xy-plane. If the bond potential energy contains a off-diagonal term $\delta U = kxy$. This term has no distinct behavior with respect to coordinate transformation $(x \to -x, y \to -y)$: it is odd $\delta U(-x, y, z) = -\delta U(x, y, z)$ for $(x \to -x)$ and even $\delta U(-x, -y, z) = \delta U(x, y, z)$ for $(x \to -x, y \to -y)$. It follows that internal torques can only be produced in molecules *without* inversion symmetry.

We introduce polar coordinates (r, θ) in the xy-plane and write $x = r\cos\theta$ and $y = r\sin\theta$ and find $\delta U = (\kappa/2)r^2 \sin(2\theta)$. The

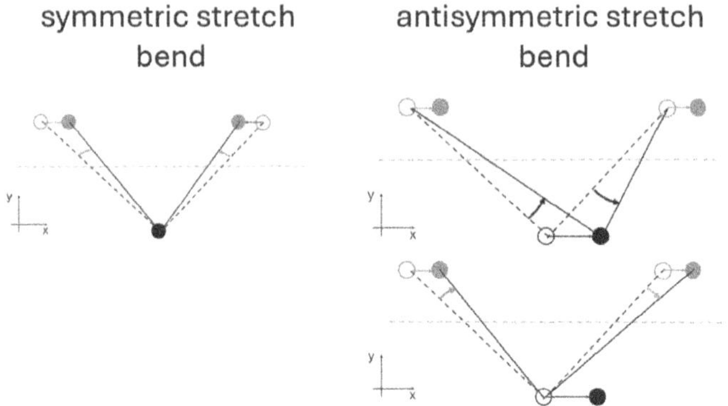

FIGURE 1.18
The coupling between the symmetric (antisymmetric) stretch and bend modes for CO_2.

torque follows $\tau = -\partial \delta U/\partial \theta = -\kappa r^2 \cos(2\theta)$ and thus varies between clockwise ($\tau < 0$) and counterclockwise ($\tau > 0$). The angular dependence of the potential $\delta U \sim \sin(2\theta)$ has maxima at $\theta_{\text{max}} = \pi/4, 5\pi/4$ and minima at $\theta_{\text{min}} = 3\pi/4, 7\pi/4$ so that the system would undergo (small) oscillations around the minima directions. We note, however, that the coupling two perpendicular modes does not produce an overall rotation in either clock- or counterclockwise direction for a coupled oscillator system without a driving force (a so-called *autonomous* system).

Molecules without inversion symmetry: The unique feature of systems without inversion symmetry is the coupling between vibrational modes along perpendicular coordinate axes (say x- and y-axis) with frequencies ω_1 and ω_2. We assume that the (dimensionless) damping constants γ_1 and γ_2, and find the equations of motion

$$\ddot{x} + \omega_1^2 x + \frac{\gamma_1}{m_1}\dot{x} + \frac{\kappa}{m_1} y = 0, \tag{1.123}$$

$$\ddot{y} + \omega_2^2 y + \frac{\gamma_2}{m_2}\dot{y} + \frac{\kappa}{m_2} x = 0. \tag{1.124}$$

In the absence of an external forcing, the trajectory in the (x, y)-plane is open (except when ω_1/ω_2 is rational) that resembles some

Lissajous-figure with decreasing amplitude (at least for small coupling $\kappa \to 0$). We note that the autonomous system does not define a rotational orientation; we say that the autonomous system is symmetric with no orientational preference.[4]

If the two oscillators are subject to external forcing $F(t) = F_0 \cos(\omega t)$ with frequency $\omega \neq \omega_x, \omega_y$, the two coordinates will eventually undergo periodic motions with frequency ω; that is, the trajectory is closed and has the 'shape' of an ellipse (or a circle). In particular, the trajectory defines either a *counterclockwise* or a *clockwise* rotation around the z-axis (out-of-the-page). We conclude that the forcing breaks the orientational symmetry.

Here, we consider the case when the forcing is aligned with one oscillator only; we choose the oscillator along the x-axis, and arrive at the set of coupled equations. For simplicity, we assume $m_1 = m_2$ and $\gamma_1 = \gamma_2$ and write $\gamma_i/m_i \to \gamma$, $\kappa/m_i \to \kappa$ and $F_0/m_1 \to F_0$. We thus have the coupled equations

$$\ddot{x} + \omega_x^2 x + \gamma\dot{x} + \kappa y = F_0 \cos(\omega t), \tag{1.125}$$

$$\ddot{y} + \omega_y^2 y + \gamma\dot{y} + \kappa x = 0. \tag{1.126}$$

We further assume that the coupling is 'weak' and only consider the lowest order approximation: the term κy can be ignored for the oscillator along x, we find

$$\ddot{x} + \omega_x^2 x + \gamma\dot{x} = F_0 \cos(\omega t), \tag{1.127}$$

$$\ddot{y} + \omega_y^2 y + \gamma\dot{y} = -\kappa x. \tag{1.128}$$

That is, the term $-\kappa x$ acts as the external force acting on oscillator along the y-axis. The (formal) solutions can be written in terms of dynamic susceptibilities:

$$x(t) = \omega\chi_x(\omega) F_0 \cos(\omega t), \tag{1.129}$$

$$y(t) = -\chi_y(\omega) x(t) = -\kappa\chi_y(\omega)\chi_x(\omega) F_0 \cos(\omega t). \tag{1.130}$$

Forced oscillator along the x-axis is in the standard form discussed in Sect. 1.1. On the other hand, the forced oscillator along the y-axis is determined by the product of two dynamic susceptibilities

$$\begin{aligned}\chi_y(\omega)\chi_x(\omega) &= [\mathrm{Re}\chi_y(\omega) + i\mathrm{Im}\ \chi_y(\omega)] \cdot [\mathrm{Re}\chi_x(\omega) + i\mathrm{Im}\ \chi_x(\omega)] \\ &= [\mathrm{Re}\chi_y(\omega) \cdot \mathrm{Re}\chi_x(\omega) - \mathrm{Im}\chi_y(\omega) \cdot \mathrm{Im}\chi_x(\omega)] \\ &\quad + i\,[\mathrm{Re}\chi_y(\omega) \cdot \mathrm{Im}\chi_x(\omega) + \mathrm{Im}\chi_y(\omega) \cdot \mathrm{Re}\chi_x(\omega)]\end{aligned} \tag{1.131}$$

[4]Our discussion of the energy transfer follows D. H. Zanette, *Energy exchange between coupled mechanical oscillators: linear regimes*, J. Phys. Comm. **2**, 095015 (2018).

TABLE 1.2
Coupling of modes

ω_x	ω_y	A_x	ϕ_x	A_y	ϕ_y	Orientation
0.4	0.4	0.766	0.872	0.586	−1.397	−
	0.8			0.720	−1.044	−
	1.2			0.701	−0.284	−
	1.6			0.413	0.302	−
0.8	0.4	0.941	1.225	0.720	−1.044	−
	0.8			0.885	−0.691	−
	1.2			0.861	0.069	−
	1.6			0.508	0.655	−
1.2	0.4	0.915	−1.156	0.701	−0.284	+
	0.8			0.861	0.069	+
	1.2			0.838	0.829	+
	1.6			0.494	1.415	+
1.6	0.4	0.540	−0.570	0.413	0.302	+
	0.8			0.508	0.655	+
	1.2			0.494	1.415	+
	1.6			0.291	−1.140	−

Some illustrative trajectories are shown in Fig. 1.19; depending on the choice of parameters, one finds either a clockwise or counterclockwise rotation, that is, a handedness.

The important fact is that forcing destroys the orientational symmetry and locks the system into either a *counterclockwise* or clockwise rotation; which orientation the system selects is not important for the purpose of this book.

1.6 Quantum Mechanics

The dynamics of particle with mass m (in three spatial dimensions) moving in a potential $V = V(\vec{r})$ is described by Newton's laws $md^2\vec{r}/dt^2 = -\nabla V$; this is second-order (ordinary) differential equation. This can be written as a system of coupled first-order differential equations. To this end, one defines the momentum $\vec{p} = md\vec{r}/dt$ and introduces the Hamilton function $H(\vec{r}, \vec{p}) = (2m)^{-1}\vec{p}^2 + V(\vec{r})$. This Hamiltonian formalism is particularly relevant for the *quantum-mechanical* description of the dynamics for atoms (ions) and (valence) electrons.

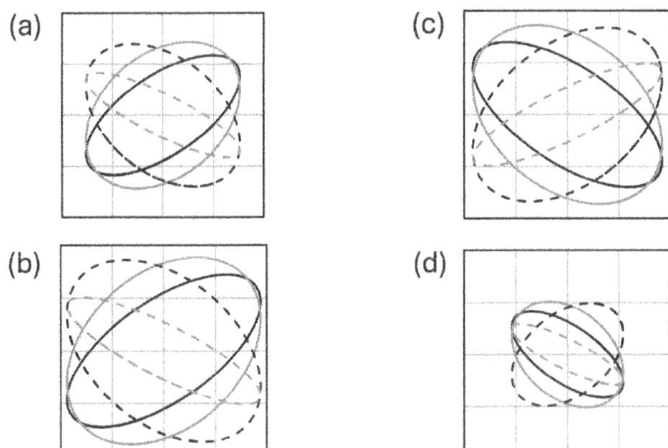

FIGURE 1.19
The trajectory of the coupled oscillator system, cf. Eqs. (1.125) and (1.26) in the $(x, y/\kappa)$-plane for $F_0 = 1$, $\omega = 1$, $\gamma = 1$. solid-black: $\omega_y = 0.4$, solid-gray $\omega_y = 0.8$, dashed-black $\omega_y = 1.4$, and dashed-gray $\omega_y = 1.6$. **a)** $\omega_x = 0.4$, **b)** $\omega_x = 0.8$, **c)** $\omega_x = 1.2$, and **d)** $\omega_x = 1.6$.

In its perhaps most familiar form, quantum mechanics is described in terms of *wave mechanics*. While the mathematical description is quite abstract, it helps to understand the underlying 'structure'. The notion of 'state' ('column vector' or 'ket') $|\psi\rangle$ and the 'linear functional ('row vector' or 'bra') $\langle r|$ are introduced so that the wave function is written $\psi(r) = \langle r|\psi\rangle$, which, in general, is complex valued. The square of the wave function describes the probability density $|\psi|^2 = \psi^*\psi$.

Central to the wavelike nature of matter is the complex-valued of wavefunctions as it explains interference (e.g., cancellations). We consider the superposition of two states $\psi = \psi_1 + \psi_2$ that the probability density follows

$$|\psi|^2 = \left[|\psi_1|^2 + |\psi_2|^2\right] + \left[\psi_1^*\psi_2 + \psi_2^*\psi_1\right]. \tag{1.132}$$

Here, the expression in the first bracket is the sum of the probability densities and captures the properties of the corresponding classical system. The second term with the mixed terms describes interference between the two states.

The overlap of two different states $|\psi\rangle$ and $|\phi\rangle$ follows

$$\langle\phi|\psi\rangle = \int d\vec{r}\,\phi^*(\vec{r})\psi(\vec{r}). \tag{1.133}$$

If the states are orthogonal $\langle\phi|\psi\rangle = 0$, the two states are uncoupled. The value of a physical quantity O (e.g., the Hamiltonian H for the energy, momentum p, etc.) corresponds to the expectation value of the corresponding operator O

$$O \longrightarrow \langle\psi|O|\psi\rangle\,. \tag{1.134}$$

The momentum (operator) is written in terms of a (spatial) gradient $\hat{p} \to (\hbar/i)\nabla$.[5] One arrives at the Schrödinger equation, $H\psi = (\hbar/i)\partial\psi/\partial t$, or

$$\left[-\frac{\hbar^2}{2m}\nabla^2 + V(\vec{r}\right]\psi(\vec{r},t) = \frac{\hbar}{i}\frac{\partial\psi(\vec{r},t)}{\partial t}. \tag{1.135}$$

We write $\psi(\vec{r},t) = \exp[i(E/\hbar)t]\psi(\vec{r},t)$ and arrive at the time-indendent Schrödinger equation

$$\left[-\frac{\hbar^2}{2m}\nabla^2 + V(\vec{r}\right]\psi(\vec{r}) = E\psi(\vec{r}) \tag{1.136}$$

subject to appropriate boundary conditions. For arbitrary value of E, Eq. (1.136) does not have a solution. The time-independent Schrödinger equation defines eigenvalues $\{E_\mu\}_\mu$,

$$E\psi_\mu = E_\mu\psi_\mu. \tag{1.137}$$

The collection of eigenvalues is referred to as the spectrum of the quantum mechanical system and corresponds to the standing waves of vibrating string (e.g., for a violin) or an air column (e.g., for an organ pipe). The Hamiltonian is Hermitian; it follows the different eigenstates are orthogonal $\langle\psi_{\mu'}|\psi_\mu\rangle = \delta_{\mu'\mu}$.

While it is certainly correct that continuous values of the energy in classical mechanics correspond to discrete values in quantum mechanics, the true impact is in fact more significant. The history of quantum mechanics is closely tight to understanding the hydrogen atom. The classical view of an electron orbiting the (stationary)

[5]This relates to the de Broglie wavelength $\lambda = h/p$ or $p = \hbar k$ with $\hbar = h/2\pi$ is the (reduced) Planck's constant.

proton in circular (or rather elliptical) orbits is flawed. The electron experiences (centripetal) acceleration and thus radiates power so that the electron would fall into the proton: the hydrogen atom is *unstable* in classical mechanics. The stationarity in the quantum mechanics, that is, discrete nature of the energy eigenvalues, implies that the electron does not radiate power. This 'suppression of power loss' has no analog in classical mechanics. There are other examples of physical phenomena that have no explanation in classical phenomena. Tunneling of particles through a potential wall is a well-known example: it explains how an α-particle can be emitted from heavy nucleus during a radioactive decay.[6]

The essential feature of quantum mechanics, that is, the wave-like nature, is the superposition of paths in coordinate-momentum, or 'phase-space.' A path defines the action $S = \int p dx$, which yields the phase $\exp(iS/\hbar)$. This is basis of the path integral formulation of quantum mechanics.[7] The phase-space origin of phases in quantum-mechanics implies that the interaction is with the potentials $(\phi, \vec{A})$ and not the fields $(\vec{E}, \vec{B})$.[8] The Hamiltonian of a charged particle follows

$$\left[\frac{1}{2m}\left(\frac{\hbar}{i}\nabla + \frac{e}{c}\vec{A}(\vec{r},t)\right)^2 + \phi(\vec{r},t)\right]\psi(\vec{r},t) = \frac{\hbar}{i}\frac{\partial\psi(\vec{r},t)}{\partial t}. \quad (1.138)$$

For time-independent fields and thus potentials, the time-independent Schrödinger equation defines stationary states $\psi(\vec{r},t) = \psi(\vec{r})$ with corresponding energy values $E = E(\phi, \vec{A})$. If the electromagnetic field is an incident electromagnetic wave, the Lorentz gauge is convenient so that $\phi = 0$ and the fields are derived from the vector potential: $\vec{E} = -\partial\vec{A}/\partial t$ and $\vec{B} = \nabla \times \vec{A}$. The Schrödinger equation,

$$\left[\frac{1}{2m}\left(\frac{\hbar}{i}\nabla + \frac{e}{c}\vec{A}(\vec{r},t)\right)^2\right]\psi(\vec{r},t) = \frac{\hbar}{i}\frac{\partial\psi(\vec{r},t)}{\partial t}. \quad (1.139)$$

[6]Tunneling is explained in *Modern Physics* texts, for example Eisberg and Resnick, Sect. 6.5 and 6.6.

[7]The classical action $S = \int p dx$ has unit (in the SI system) $[S] = \mathrm{kg\,m^2/s}$, which is the unit of angular momentum $[S] = [L]$. It is not surprising in this light that the simple Bohr model for the hydrogen atom actually quantizes the angular momentum of the orbiting electron.

[8]A striking example is the propagation of charged particle in a region that includes a magnetic field contained in a solenoid. That is, the particle does not 'feel' the magnetic field, but the phase is changed. This is the Aharonov-Bohm effect; an elementary discussion is presented in Feynman Vol. II—Chapter 15.

1.7 Annotated Bibliography

Mathematics: Books on 'Mathematical Methods' are numerous; a standard text is G. B. Arfken, H. J. Weber and F. E. Harris, *Mathematical Methods for Physicists: A Comprehensive Guide* 7th Ed. (Academic Press, Cambridge, MA, 2012).

Complex-rotation: The relationship between complex numbers and rotation is discussed at an elementary level in R. Arianhod, *Vector—A Surprising Story of Space, Time, and Mathematical Transformation* (The University of Chicago Press, Chicago and London, 2024).

Electricity and magnetism: The standard undergraduate texts are D. J. Griffiths, *Introduction to Electrodynamics* (Cambridge University Press, New York, 2023) and E. M. Purcell and D. J. Morin, *Electricity and Magnetism* (Cambridge University Press, New York, 2013). At the graduate level, we have J. D. Jackson, *Electrodynamics* (J. Wiley & Sons, New York, 1999) and A. Zangwill, *Modern Electrodynamics* (Cambridge University Press, New York, 2012).

Units: A thorough discussion of MKS and cgs units can be found in Purcell and Morin. Jackson made the switch from cgs to MKS from the second to the third edition.

Optics: Electromagnetic waves are described in texts on Electricity and Magnetics and Optics texts. The standard undergraduate text is E. Hecht, *Optics* 5th Ed. (Pearsons, Boston, 2021) and the standard graduate text is E. Wolf and M. Born, *Principles of Optics: Electromagnetic Theory of Propagation Interference and Diffraction Light* (Pergamon, London, 1980).

Oscillations: The basics of oscillations are covered in detail in introductory texts, for example, D. Halliday, R. Resnick and J. Walker, *Fundamentals of Physics* 7th Ed. (J. Wiley & Sons, New York, 2005). Coupled oscillations are discussed in advanced undergraduate texts in classical mechanics, see, for example, S. T. Thornton and J. B. Marion, *Classical Dynamics of Particles and Systems* (Brooks/Cole Cengage, Boston, MA, 2008).

Vibrational modes in molecules: Normal modes in molecules are discussed in detail in chemistry books, see, for example, R. J. Silbey, R. A. Alberty, M. G. Bawendi, *Physical Chemistry* 4th Ed. (J. Wiley & Sons, 2005); Chapter 13.

Quantum mechanics: A thorough but non-technical description is given in R. P. Feynman, R. B Leighton, and M. Sands *Lectures on Physics* Vol. III (Addison-Wesley, Reading, MA, 1965). Elementary quantum mechanics beyond the simple Bohr model is presented in Modern Physics text; a standard reference is R. Eisberg and R. Resnick, *Quantum Physics of Atoms, Molecules, Solids, Nuclei, and Particles* 2nd Ed. (J. Wiley & Sons, New York, 1985). A standard undergraduate text is D. J. Griffiths, *Introduction to Quantum Mechanics* (Prentice Hall, Upper Saddle River, NJ, 1995) and standard graduate texts are J. J. Sakurai, *Modern Quantum Mechanics* (Addison-Wesley, Redwood City, CA, 1985) and S. Weinberg, *Lectures on Quantum Mechanics* (Cambridge University Press, New York, 2013).

Path integrals: The role of action in classical mechanics is discussed in R. P. Feynman, R. B Leighton, and M. Sands *Lectures on Physics* Vol. II (Addison-Wesley, Reading, MA, 1965); Chapter 19: *The Principle of Least Action.* The standard reference for the path-integral formulation of quantum mechanics is R. P. Feynman and A. R. Hibbs (emended by D. F. Styer), *Quantum Mechanics and Path Integrals* (Dover, Mineola, NY, 2005). A brief (but very useful) discussion is presented in G. Baym, *Lectures on Quantum Mechanics* (Benjamin/Cummings, Reading, MA, 1969), Chapter 3.

2

Dielectric medium

Optics (such as reflection and refraction) refers to the response of dielectric matter to a propagating electromagnetic wave. Matter becomes polarized $\vec{P} \neq 0$. The term *linear* optics refers to the case when the polarization increases linearly with an applied electric field $P = \chi E$ so that dielectric susceptibility is independent of the electric field $\chi(E) = \text{const}$. This case applies to light with intensity equal to that of (bright) sunlight or less, $|E| \leq 10^3\,\text{V/m}$; this limit is referred to as *linear* optics. It follows that susceptibility represents properties of individual molecules and cooperative effects between neighboring molecules are irrelevant.

Since the size of molecules is much smaller than the wavelength of visible light, an individual molecule can be treated as a collection of point dipoles Generally, molecular vibrations are 'IR' active so that the relevant vibrational frequencies are much smaller than frequencies corresponding to visible light. That is, the response of molecules to the incident light is 'off-resonance,' however, a (small) frequency dependence remains so that the electric susceptibility depends on the frequency of the incident light $\chi = \chi(\omega)$. Consequently, the index of refraction is (slightly) different for the different colors compromising white (sun-)light: this is referred to as (frequency-) dispersion.

In Sect. 2.1, we use estimates of dielectric properties of liquid water to understand how linear response $P(E) = \chi E$ is consistent with an independent response of each molecule to an external electric field. Dipole-dipole interactions of neighboring molecules, or cooperative phenomena are irrelevant for linear optics, and thus for optical activity, and are only relevant for nonlinear effects. We introduce the general description in Sect. 2.2. In Sect. 2.3, we discuss the orientation of polar molecules in an external electric field. The basic properties of induced dipole moments are presented in Sect. 2.4. We discuss the frequency dispersion in Sect. 2.5.

DOI: 10.1201/9781003560944-2

2.1 Linear Dielectrics

The 'typical' permanent dipole moment of a molecule is *one debye* $p \simeq 1\,\mathrm{D} = 10^{-30}\,\mathrm{C\,m}$. For a 'typical' distance between neighboring atoms or molecules $\mathcal{L}$. We found that the 'typical' energy associated with the alignment of neighboring dipoles is of the same order or smaller than the thermal energy $(4\pi\epsilon_0)^{-1}p^2/r^3 \leq k_B T$; That is, any local ordering of nearest-neighbor dipoles is destroyed by thermal fluctuations. As a result, individual molecules undergo tumbling motion on the time scale $\mathcal{T}$ so that their average dipole moment averages to zero $\langle p\rangle_{\mathcal{T}} \simeq 0$.

An alignment of dipole moment requires an external electric field such that the interaction energy is several order of magnitudes larger than thermal energy. We set $U = pE \simeq 100\mathcal{U}$ and find the estimate for the static (or DC) electric field

$$E_{\mathrm{DC}} > \frac{100\,\mathcal{U}}{1\,\mathrm{D}} = \frac{10^{-19}\,\mathrm{J}}{10^{-30}\,\mathrm{C\,m}} = 10^{11}\,\frac{\mathrm{V}}{\mathrm{m}}, \tag{2.1}$$

which is consistent with the electric field associated with the intensity of laser light. In turn, a 'locked' dipole moment $p = 1\,\mathrm{D}$ produces a constant electric field at the location of its nearest neighbor; for $r = 2\mathcal{L}$, we find the estimate for the magnitude $U_{\mathrm{nn}} \simeq (4\pi\epsilon_0)^{-1}p/r^3 = 10^{10}\,(\mathrm{N\,m^2/C^2}) \cdot 10^{-30}\,\mathrm{Cm}/(10^{-10}\,\mathrm{m})^3 = 10^{10}\,\mathrm{V/m}$ so that $U_{\mathrm{nn}} \simeq U_{\mathrm{DC}}/10$. We expect that the summation of the electric fields produced by all its neighbors produces a local field that is equal or greater than the DC electric field $E_{\mathrm{local}} > E_{\mathrm{DC}}$. We conclude that the view of dielectric matter from introductory texts is misleading. The alignment of dipole moment requires a DC electric field that is found only in laser beams; furthermore, it is inconsistent with the presumed linear behavior of the polarization of matter.

In fact, a DC electric field with a magnitude corresponding to that of (bright) sunlight $E \simeq 10^3\,\mathrm{V/m}$ yields an interaction energy

$$U \simeq 10^{-30}\,\mathrm{C\,m} \cdot 10^3\,\frac{\mathrm{V}}{\mathrm{m}} = 10^{-27}\,\mathrm{J}, \tag{2.2}$$

which is very small compared to the characteristic energy

$$U \simeq 10^{-6}\mathcal{U}. \tag{2.3}$$

We conclude that in this case, the strength of the external field is too small to significantly affect the tumbling motion of the molecules. The molecules still point in all directions with only a very small 'preference' for alignment in the direction of the DC field. That is, the polarization requires a statistical treatment such that the *average* dipole moment is non-zero and points in the direction of the DC field, $\langle\vec{p}\rangle \propto \vec{E}$. We expect that the electric field produced by the average dipole moment at the location of the nearest neighbor is negligibly small compared to the DC field $E_{\mathrm{nn}} = (4\pi\epsilon_0)^{-1}\langle p\rangle/r^3 << E_{\mathrm{DC}}$. It follows that the dipole moments are independent of each other and the polarization is proportional to the DC electric field, $\vec{p} \sim \vec{E}$. That is, we recover the assumptions underlying *linear* optics.

While we discussed a particular case (namely liquid water), a clear picture of dielectric matter emerges that is different from that described in elementary texts. Dipole moments of the order of 1 debye (1 D), associated with permanent molecular dipole, are extraordinarily large that would produce a strong internal electric field if the dipole moments were aligned in roughly the same direction. Dielectric properties of matter under ordinary conditions (e.g., sunlight) is explained by dipole moments that are many orders of magnitude smaller. We arrive at two scenarios. (1) For molecules with permanent dipole moments, the alignment of dipoles is destroyed by thermal fluctuations (Brownian motion) so that the average dipole moment approaches zero at high temperatures $p \to \langle p\rangle \simeq 0$ for $T \to \infty$. (2) For atoms and molecules without permanent dipole moment, their structural properties lead to a stiffness such that the centers of negative and positive charges are separated by an exceedingly small amount so that the resulting dipole moment is much smaller than 1 debye.

2.2 General Description

We have so far limited our discussion to the static limit $\omega \to 0$. We are generally interested in time-dependent behavior $\omega > 0$ so that the time-averaged dipole moment is not a useful starting point for a description of dielectric matter. We note that polarization is macroscopic quantity; that is, it is the sum of elementary dipoles in a parcel with volume ΔV at $\vec{r}$,

$$\vec{P} = \frac{1}{\Delta V_{\rm p}} \sum_i \vec{p}_i. \tag{2.4}$$

This is the starting point for a formal discussion of dielectric matter.

This can be written in terms of a product of an average dipole moment $\langle \vec{p} \rangle$ and the number density n

$$\vec{P} = n \langle \vec{p} \rangle . \tag{2.5}$$

We find the number density for condensed phases (i.e., liquids and solids) and gases. (1) For liquids and solids (i.e., condensed phases), we find an estimate using values corresponding to (liquid) water H_2O with molar mass $M = 18\,\mathrm{g/mol}$. The mass density is $\rho = 10^3\,\mathrm{kg/m^3}$ that corresponds to the number density $n_{\rm c} = (10^6\,\mathrm{g\,m^{-3}})/(18\,\mathrm{g\,mol^{-1}}) \cdot 6 \times 10^{23}\,\mathrm{mol^{-1}} = 3.3 \times 10^{28}\,\mathrm{m^{-3}}$. (2) For gases (e.g., N_2 and O_2 with molar masses $M = 28\,\mathrm{g/mol}$ and $M = 32\,\mathrm{g/mol}$, respectively), the mass density $\rho \simeq 1\,\mathrm{kg/m^3}$ corresponds to the number density $n_{\rm g} = (10^3\,\mathrm{g\,m^{-3}})/(30\,\mathrm{g\,mol^{-1}}) \cdot 6 \times 10^{23}\,\mathrm{mol^{-1}} = 2 \times 10^{25}\,\mathrm{m^{-3}}$.

In linear response regime, we write the polarization in terms of the polarizability γ,

$$\vec{p} = \gamma \epsilon_0 \vec{E}. \tag{2.6}$$

We then have

$$P = \gamma n \epsilon_0 E. \tag{2.7}$$

2.3 Polar Molecules

The potential energy of a permanent dipole in a constant electric field is $U = -p_0 E \cos\theta$, where θ is the angle between the electric field and the (instantaneous) dipole moment. The lowest (highest) energy corresponds to case when the dipole moment is parallel (antiparallel) to the electric field. That is, at zero temperature, the dipole moments are aligned with the external field. In thermal equilibrium, the alignment is not perfect and the probability of an angle θ is given by the Boltzmann factor

$$\xi(\theta) = \frac{1}{4\pi} \exp\left(-\frac{(-p_0 E \cos\theta)}{k_B T} \right). \tag{2.8}$$

For an electric field $E = 10^3\,\mathrm{V/m}$, we have $pE = 10^{-28}\,\mathrm{C\,m} \cdot 10^3\ \mathrm{V/m} = 10^{-25}\,\mathrm{J} = 10^{-4}\,\mathcal{U}$. Since $k_BT = 2.5\mathcal{U}$ at room temperatures, we have $U/k_BT << 1$ so that we can use the approximation

$$\xi(\theta) \simeq \frac{1}{4\pi}\left(1 + \frac{pE\cos\theta}{k_BT}\right). \tag{2.9}$$

Thus, the average dipole moment follows by integrating with respect to the azimuthal angle $0 \leq \phi \leq 2\pi$ and polar angle $0 \leq \theta \leq \pi$,

$$\langle p\rangle_T = \int_0^{2\pi} d\phi \int_0^{\pi} d\theta\, \sin\theta p_0 \cos\theta \cdot \xi(\theta) = \frac{p_0^2 E}{3k_BT}, \tag{2.10}$$

where we used $\int_0^{\pi} \sin\theta \cdot \cos^2\theta = 2/3$.

We get a 'feel' for $\langle p\rangle_T$ by considering 'typical' values. We choose $E = 10^3\,\mathrm{V/m}$ and write $\langle p\rangle_T = \zeta p_0$ with $\zeta = p_0E/3k_BT = 10^{-4}\mathcal{U}/(2.5\mathcal{U}) = 10^{-5}$. That is, if a parcel contains 10^5 polar molecules, all but one are directed in perfectly random direction such that the average average dipole momenta vanish: this corresponds to the high temperature limit $\langle p\rangle_{T\to\infty}$. At room temperature, only one molecule is aligned with the electric field, while the vast majority is directed in random direction.

We find the volume of 10^5 from the number density $\delta V = 10^5/(3.3\times10^{28}\,\mathrm{m}^3) = 3\cdot10^{-24}$. The size of the parcel must be larger to suppress fluctuations. Assuming a 3% (0.03) fluctuations, the law of large numbers from statistics) suggests $\Delta V = (0.03)^{-2}\delta V = 10^{-21}\,\mathrm{m}^3$. We arrive at an estimate for the size of a parcel $\Delta V = L^3$ with $L = 10^{-7}\,\mathrm{m} = 100\,\mathrm{nm}$.

The polarization follows

$$P = \frac{np_0^2E}{3k_BT}, \tag{2.11}$$

Since $P = \chi_e\epsilon_0E$, we find electric susceptibility of water $\chi_e = P/(\epsilon_0E)$

$$\chi_e = \frac{np_0^2}{3k_BT\epsilon_0}. \tag{2.12}$$

We check the order of magnitude

$$\chi_e = \frac{3\times10^{28}\cdot(10^{-30}\,\mathrm{C\,m})^2}{4\times10^{-21}\,\mathrm{J}\cdot9\times10^{-11}\,\mathrm{C}^2/(\mathrm{N\,m}^2)} \simeq 10^{-1}, \tag{2.13}$$

which is a reasonable value.

2.4 Induced Dipole Moments

The induced dipole moments for atoms and molecules have different physical origins: for atoms, it relates to displacement of the electron 'cloud,' while for molecules, it relates to the stretching or compression of bond between atoms. We discuss these two different mechanisms separately.

Atoms: While the nucleus can be described classically (i.e., Newtonian physics), electrons require a quantum mechanical description. Here we are only interested in the order of magnitude, a semi-classical description of electrons is sufficient.

In the absence of an external electric field, the negatively charged electrons orbit with radius r_0 the positively charged nucleus in a circular orbit. The centers of the negative and positive charges coincide and the permanent dipole moment is zero, cf. Fig. 2.1. In the presence of an external electric field, the nucleus moves in the direction of the electric field, whereas the electron orbit is distorted in the direction opposite to the electric field. More specifically, the center of the orbit moves in the direction opposite to the electric field. The relative displacement of the electron from the nucleus is described by a change in the radius $r = r_0 + \delta r$; the stability of the electron orbit implies an 'effective' restoring force that balances the electric force $eE \sim k_e \delta r$ with $k_e \simeq 10\,\mathrm{N/m}$. The displacement of the centers of positive and negative charges follows $\delta r = (e/k_e) \cdot E$.

For a numerical value, we choose an electric field $E = 10^3\,\mathrm{V/m}$, we find $\delta r = [1.6 \times 10^{-19}\,\mathrm{C}/(10\,\mathrm{N/m}) \cdot 10^3\,\mathrm{N/C} \simeq 10^{-17}\,\mathrm{m}$. We calculate the relative change $\delta r / r_0 \simeq 10^{-17}\,\mathrm{m}/10^{-10} = 10^{-7}$, which is a *tiny* fraction (*one-part-in-10-million*) of the radius of the electron orbit. The induced dipole moment follows $p = (e^2/k_e)E$ with $e^2/k_e \simeq (1.6 \times 10^{-19}\,\mathrm{C})^2/k_e = 2.5 \times 10^{-39}\,\mathrm{C\,m}$, or $p_{\mathrm{atom}} \simeq 10^{-9}\,\mathrm{D}$.

In condensed phases, this dipole would lead to a polarization

$$P = 3.3 \times 10^{28}\,\mathrm{m}^{-3} \cdot 10^{-39}\,\mathrm{C\,m} = 3.3 \times 10^{-11}\,\frac{\mathrm{C}}{\mathrm{m}^2}. \tag{2.14}$$

Since $\epsilon_0 E \simeq 10^{-10}\,\mathrm{C}^2/(\mathrm{N\,m}^2) \cdot 10^2\,\mathrm{N/C} = 10^{-8}\,\mathrm{C}^2/\mathrm{m}^2$. The electric susceptibility follows

$$\chi_e = \frac{P}{\epsilon_0 E} = \frac{3.3 \times 10^{-11}\,\mathrm{C/m}^2}{10^{-8}\,\mathrm{C/m}^2} \simeq 10^{-3}. \tag{2.15}$$

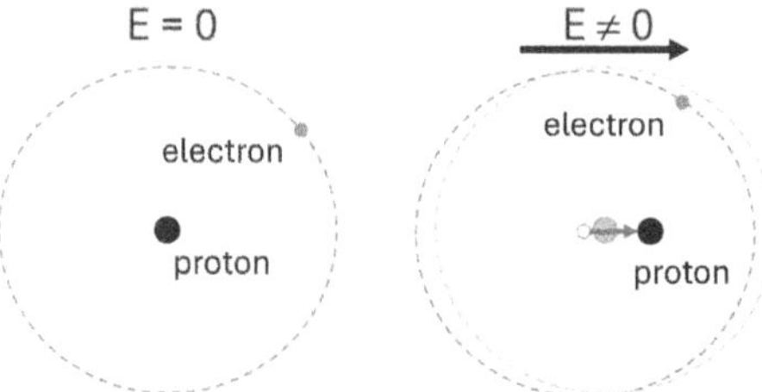

FIGURE 2.1
In a hydrogen atom, the electron orbits a proton on a circular orbit. (a) For zero electric field $E = 0$, the centers of the positive and negative charges coincide so that the dipole moment is zero. (b) In the presence of an electric field $E \neq 0$, the centers of the electron orbit and the proton move in opposite directions, creating a non-zero dipole moment.

In the gas phase, the number density is about a factor 10^{-3} smaller so that we find $\chi_e \simeq 10^{-6}$. This is consistent with a negligible relative electric constant in the gas phase $\epsilon_r \simeq 1$; that is, the dielectric properties can be ignored and gases can be treated as 'vacuum.'

Molecules: We consider nonpolar molecules, that is, molecules with no permanent dipole moment in the absence of an electric field. The centers of negative and positive charge distributions coincide. We use the linear CO_2-molecule as an illustrative example, cf. Figs. 2.2 and 2.3. The oxygen molecules are *negatively* charged $O^{-\delta}$ and the carbon atom is *positively* charged $C^{+2\delta}$ with $\delta = 0.73$; the bond length is $a = 116.3$ pm. The molecule consists of dipole moments $\vec{p}_{1,0}$ and $p_{2,0}$ corresponding to the C-O and O-C bonds, respectively. We have $|\vec{p}_{1,0}| = |\vec{p}_{2,0}| = 3.0\,\mathrm{D}$. The two dipole moments cancel out each other so that the net dipole moment vanishes $\vec{p}_0 = \vec{p}_{1,0} + \vec{p}_{2,0} = 0$.

In general, a non-zero external electric field has components both parallel (along the x-axis) and perpendicular (along the y- or z-axis) to the molecular axis. If the electric field is directed along the molecule, the oxygen atoms move along the $-x$-axis and the carbon atom moves along the $+x$-axis. In the presence of an electric field $F = eE$, one C-O bond is stretched, while the other is squeezed $a \to a \pm \delta a$; that is, the molecular vibration associated with the electric field *excites* corresponds to the *antisymmetric*

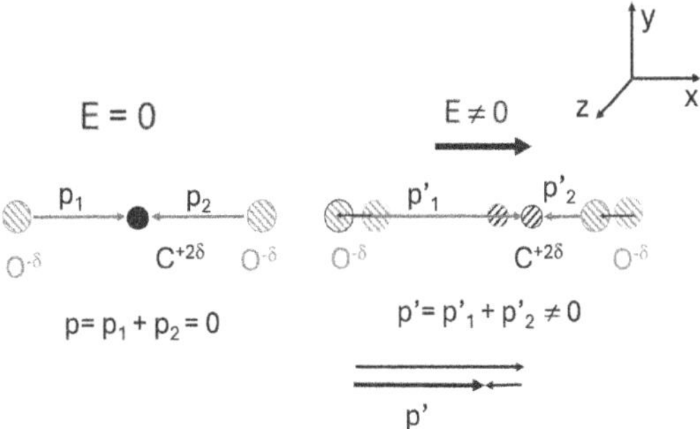

FIGURE 2.2
The polarization of a CO_2 molecule due to an electric field in the direction of the molecular axis.

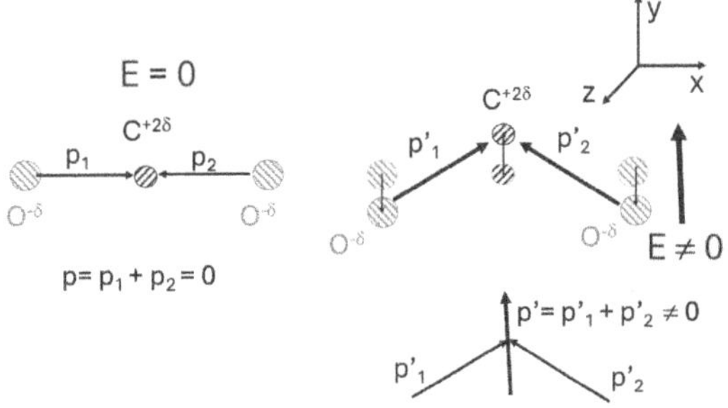

FIGURE 2.3
The polarization of a CO_2 molecule due to an electric field in a direction perpendicular to the molecular axis.

stretch. For an electric field in the (y, z)-plane, the carbon and oxygen ions move in the direction and opposite to the electric field, respectively; that is, the electric field *excites* the *bending* mode.

We estimate the change in bond length $\delta a = eE/k$ and k is the spring constant characterizing the chemical bond $k \simeq 10^2\,\mathrm{N/m}$. For an electric field $E = 10^3\,\mathrm{V/m}$, we find $\delta a \simeq (0.73 \cdot 1.6 \times 10^{-19}\,\mathrm{C}) \cdot 10^3\,\mathrm{V/m}/(10^2 \times 10^{N/m}) \simeq 10^{-18}$ so that $\delta a/a \simeq 10^{-8}$; this small change of the bond length reflects the stiffness of the chemical bond. As a result, the dipole moments are unequal $|vecp_1| = p_{1,0} \pm$

δp and with $\delta p \simeq 2(\delta \cdot e)\delta a$. We find

$$\delta p \simeq 10^{-19}\,\mathrm{C} \cdot 10^{-18}\,\mathrm{m} = 10^{-37}\,\mathrm{C\,m} = 10^{-7}\,\mathrm{D}. \tag{2.16}$$

We assume a number density corresponding to the condensed phase and find the polarization

$$P = 3.3 \times 10^{28}\,\mathrm{m}^{-3} \cdot 10^{-37}\,\mathrm{C\,m} = 10^{-9}\,\frac{\mathrm{C}}{\mathrm{m}^2}. \tag{2.17}$$

We find the induced electric susceptibility for molecules

$$\chi = \frac{10^{-9}\,\mathrm{C\,m}}{(10^{-11}\,\mathrm{C}^2/\mathrm{N\,m}^2) \cdot (10^3\,\mathrm{V/m})} = 10^{-1}. \tag{2.18}$$

Despite their differences, the induced dipole moment is determined by the ratio of the square of the elementary charge e^2 and a ‘spring’ (or elastic) constant k:

$$p \propto \frac{e^2}{k} E. \tag{2.19}$$

The exact value depends on the value of partial (or ‘effective’) charge involved in the charge redistribution and the value of the spring constant. The partial charge is typically close to unity and the spring constant is similar for atoms and molecules, $k \simeq 10^2\,\mathrm{N/m}$. Thus their induced dipole moments are quite similar; even for a reasonably strong electric field E_0, we found a value that is about *one-millionth* of a debye

$$p_{\mathrm{ind}} \simeq \frac{(10^{-19}\,\mathrm{C})^2}{10^2\,\mathrm{N/m}} \cdot 10^3\,\frac{\mathrm{V}}{\mathrm{m}} = 10^{-37}\,\mathrm{C\,m} \sim 10^{-6}\,\mathrm{D}. \tag{2.20}$$

We estimate the electric field produced by such an induced dipole moment at the site of its nearest neighbor at the distance $2\mathcal{L}$,

$$E_{\mathrm{nn}} = \frac{1}{4\pi\epsilon_0}\frac{p_{\mathrm{ind}}}{\mathcal{L}^3} \simeq 10^{10}\,\frac{.N\,\mathrm{m}}{\mathrm{C}^2} \cdot \frac{10^{-37}\,\mathrm{Cm}}{(2 \times 10^{-10}\,\mathrm{m})^3} = 10^2\,\frac{\mathrm{V}}{\mathrm{m}}; \tag{2.21}$$

that is, the internal electric field is about one-tenth of the external electric field.

We conclude that the induced polarization is characterized by a susceptibility that is proportional to the number density of atoms or molecules, the square of the elementary charge, and spring constant characterizing the response of the system:

$$\chi \simeq \frac{ne^2}{k\epsilon_0}, \tag{2.22}$$

where the dependence on ϵ_0 only applies to the MKS system. The corresponding equation in the cgs system would read $\chi = 4\pi n e^2/k$. We check the order of magnitude

$$\chi \simeq \frac{10^{28}\,\mathrm{m}^{-3} \cdot (10^{-19}\,\mathrm{C})^2}{10^2\,\mathrm{N/m} \cdot 10^{-10}\,\mathrm{C}^2/\mathrm{N\,m}^2} \simeq 10^{-2}. \tag{2.23}$$

Since the relative electric permeability determines the index of refraction $n = \sqrt{\epsilon}$ and $v = c/n$, we find a change in the propagation speed by about 5%. This is reasonable estimate.

Stretching of molecular bonds implies that the elastic potential energy is stored in molecules. In a simple scenario, we imagine that a slab of dielectric material is inserted between the two parallel plates of an uncharged capacitor. We attach a battery to the capacitor to produce the external electric field and therefore the polarization of the dielectric matter; as a result, chemical energy stored in the battery is transformed into energies associated with the electric field (inside and outside the dielectric slab) and the elastic potential energy associated with the deformation of molecules (i.e., the polarization inside the dielectric matter).

2.5 Frequency Dispersion

So far, we only discussed the response to static electric fields. We now turn our attention to the response of a polar molecule to a time-dependent electric field. The response depends on the interplay between the dynamics characterizing the dielectric matter and the external forcing. While the external electric field can have an arbitrary time-dependence, the method of Fourier transforms, cf. Sect 1.1, shows that we can limit our discussion to the response to periodically-varying electric fields. We use complex-notation $E(t) = \mathcal{E}\exp(-i\omega t)$ that matches the time-dependence of a propagating wave $E(\vec{r}, t) = E_0 \exp(i[\vec{\mathsf{k}} \cdot \vec{r} - \omega t])$ so that the complex-valued amplitude depends on the location of the molecule $\vec{r}_0$ follows $\vec{E} = \vec{E}_0 \exp(i\vec{\mathsf{k}} \cdot \vec{r}_0)$.

Electromagnetic waves vary both in time as in space. The behavior of dielectric matter is determined by properties of molecules and atoms characterized by the length scale $\mathcal{L} = 10^{-10}\,\mathrm{m}$. If we limit our discussion to wavelengths $\lambda > 10\mathcal{L} = 10^{-9}\,\mathrm{m}$, we can ignore the spatial variation and treat an incident wave as spatially homogenous but varying in time. This condition is not too restrictive as it only limit the discussion to wavelengths longer than the X-ray part of the spectrum but includes the visible part.

The time-dependent response of dielectric matter depends on characteristic time scales of molecules (and atoms). The tumbling motion of polar molecules and the vibrational motion of non-polar molecules have (roughly) the same characteristic timescale $T_0 \simeq \mathcal{T} = 10^{-11}$ s; this corresponds to a frequency $\omega_0 = 2\pi/T_0 \simeq 10^{11}$ Hz and falls in the IR part of the electromagnetic part of the spectrum. **Response of a single bond**: The stretching of a single bond due to an oscillatory forcing is described by the response function $l(\omega) \sim \chi(\omega)$. We find the induced dipole moment

$$p = \frac{eE}{m} \frac{1}{\omega_i^2 - \omega^2 - i\omega\gamma_i}. \tag{2.24}$$

We expect that the response of the dielectric matter is drastically different for low and high frequencies. For frequencies lower than IR part of the spectrum $\omega < \omega_0$, molecules 'see' a nearly static external electric field, whereas for frequency in the visible and UV part of the spectrum $\omega > \omega_0$, molecules 'see' a small average electric field $E(t) \to \langle E \rangle \simeq 0$.

In the case of circularly polarized light, the magnitude of the electric field is constant but the direction of the electric field rotates about the direction of propagation with angular frequency ω. The situation is shown in Fig. 2.4. If a circularly polarized light is incident on dielectric matter, the molecular bond is stretched by a fixed length r and rotates with angular speed ω. The corresponding moment of inertia is $I = mr^2$. The forces on the charge are the restoring spring force $kr = m\Omega^2 r$ (directed inward) and the electric force qE (directed outward). Thus, Newton's second law yields $m\omega^2 r = m\omega_0^2 r - qE$ so that the bond is stretched by a fixed distance r

$$r = \frac{eE}{m(\omega_0^2 - \omega^2)}. \tag{2.25}$$

This expression is identical to the amplitude for oscillation of the molecular bond for linearly polarized light.

Since the bond is stretched by a fixed distance and the molecule rotates with constant (angular) speed, both the kinetic energy and the elastic potential energy are constant. In particular, the total energy (sum of kinetic and potential energy) is identical for linearly

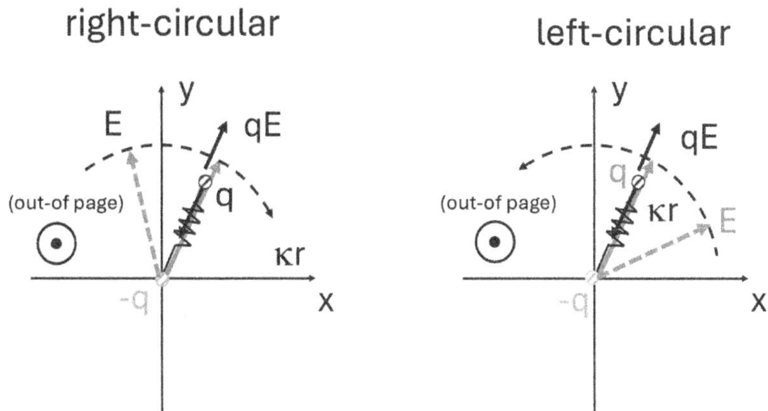

FIGURE 2.4
Stretching of bonds due to circularly polarized light. The direction of propagation is along the z-axis (out-of page). The electric field rotates clockwise and counterclockwise for right- and left-circular polarization, respectively.

and circularly polarized light. There is, however, an important difference between the two types of polarization states since the rotation of the stretched bond implies that the molecule also carries angular momentum either parallel or antiparallel to the direction of wave propagation. We have $L_z = I\omega$, or

$$L_z = \pm m \left(\frac{e^2 E}{m(\omega_0^2 - \omega^2)} \right)^2 \omega \quad \begin{cases} - & \text{right-circular} \\ + & \text{left-circular} \end{cases} . \tag{2.26}$$

Evidently, this angular momentum is transferred from the incident light to the molecule: that is, for incident light with right- (left-) circularly polarized light, the molecule rotates in clockwise (counterclockwise) direction with respect to the $+z$-axis.

In many cases, this qualitative behavior is sufficient to understand the overall frequency-dependent of dielectric matter. In the case of water, the polar molecules are aligned and produce a large polarization if the period associated with the time-dependent electric field is lower than the frequency of the tumbling dynamics; in contrast, the molecules are arranged randomly and only produce a small polarization if the period associated with the time-dependent electric field is faster than the frequency of the tumbling dynamics.

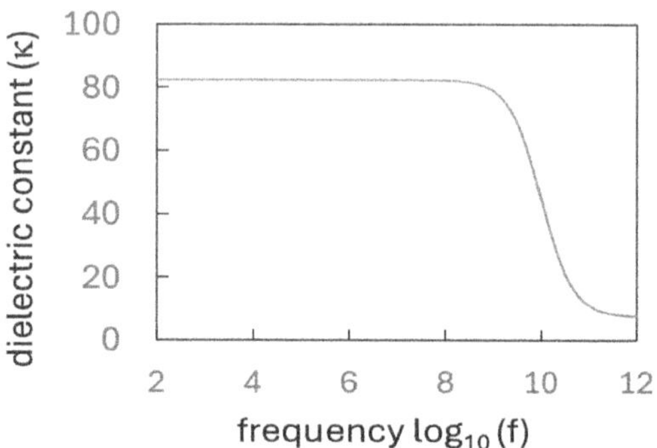

FIGURE 2.5
Index of refraction for liquid water (H_2O). The high index of refraction for low frequencies implies that a water surface acts essentially as a mirror for radiowaves; whereas the low index of refraction for high frequencies implies that the same water surface is more-or-less transparent for visible light. The transition occurs over a relatively small range of frequencies.

The response of water to an incident electromagnetic wave matches the expected frequency dependence: this index of refraction is high $n \simeq 80$ for radio frequencies $\omega \simeq 10^8$ Hz and rapidly drops in a narrow range $10^{10}\,\text{Hz} < \omega < 10^{11}\,\text{Hz}$, and then approaches the high-frequency limit for $\omega > 10^{12}$ Hz, cf. Fig. 2.5.

We consider an interface between water and air, and assume that electromagnetic wave is incident from air. A value of the index of refraction much greater than unity implies that water surface acts both as a *reflector* and *refractor*. We can see what is below the water surface and can also see the surroundings reflected on the water surface. We note that introductory physics texts may give the erroneous impression that the frequency dependence of water is weak; the discussion on (frequency-) dispersion is limited to the visible part of the spectrum so that the weak dependence reflects the small frequency range of the visible part of the spectrum.

Dielectric response of molecules: We now turn to the response of (large) non-polar molecules to an oscillating electric field. We adopt a complex notation of the (time-dependent) electric field $\vec{E} = \vec{E}_0 \exp(i\omega t)$. The interaction of light with matter involves the coupling to dynamical degrees of freedom of charged particles, that is electrons and ions. The strength of the coupling is inversely proportional to the mass. Since electron mass is small to compared to the mass of an atom (ion) ($m_e/\mathcal{M} \simeq 1/1,836$), we conclude that dielectric response involves electronic transitions from lower- to a higher-discrete electronic state.

The explanation of line spectra in terms of discrete electron 'orbits' played a central role in the development of *Quantum Mechanics* in the early 20th century. A proper treatment involves finding solutions of an appropriate time-independent Schrödinger equation. For a qualitative discussion, a semiclassical description in terms of Bohr's quantization condition is often adequate: the quantization can be expressed as the condition that de Broglie wavelength of the electron forms a standing wave on its orbit. A bigger radius of the orbit corresponds to an electron that is more loosely bound electron, that is, to a higher energy eigenvalue of the electron. This implies, in particular, that the length scale associated with electronic transitions is of the order of the size of an atom, that is, we identify the length scale associated with electronic transition with the characteristic length $\mathcal{L}$.

There is a one-to-one correspondence between a discrete absorption/emission line and a jump of the electron from one such level to another. The jumps occur on time scales much shorter than any other relevant time (e.g., the period in description of the electron orbit in a classical description): the jumps are considered as instantaneous; their likelihood are characterized in terms of Einstein coefficients; details are explained in *Modern Physics* texts.

The interactions of light with molecules still involve jumps of electrons; however, the description must be modified since electronic and atomic degrees of freedom are coupled. Such an interaction is evident from the relevant energy scales: the energies associated with electronic jumps are of the order of a few electron volts (eV), whereas the vibrational energies are of the order of a milli-electron volts. It follows that a single electronic state (e.g., the ground or an excited state) corresponds to a 'range' of atomic coordinates, each corresponding to a different elastic potential energy in the case of the coupling to vibrational degrees of freedom. The electronic energy states are superimposed with potential energy curves: the total energy (sum of electronic and vibrational energy) is plotted vs an atomic coordinate, e.g., the bond length for a diatomic molecule. The minimum potential energy corresponds to a zero dipole moment of the molecule. The electronic state changes the overall charge distribution so that the atomic configuration

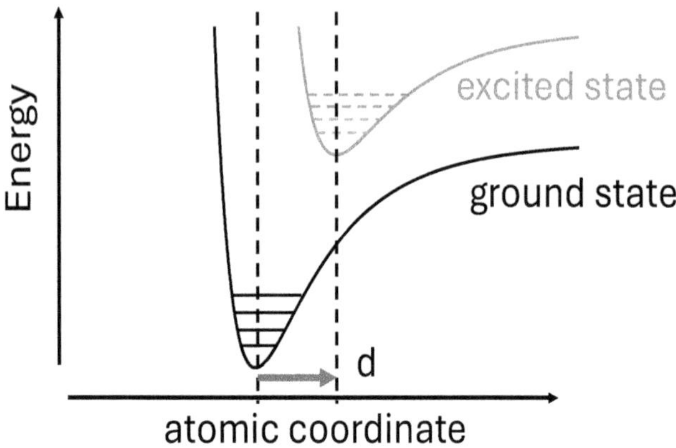

FIGURE 2.6
Potential wells and electron energy levels of the ground and excited states; the separation of the minima is associated with a displacement of atoms (ions).

corresponding to the minimum energy is different for each electronic state, cf. Fig. 2.6. An electronic transition is (essentially) instantaneous so that the atomic configuration is unchanged when the electron jumps from the ground to the first excited state, say. As a result, the atomic configuration in the excited states does not correspond to the minimum so that the electronic transition is associated with an (induced) dipole moment. For organic molecules, only smaller part of the entire molecule contributes to the structural re-arrangement associated the electronic transition: such units are referred to as *chromophores*.

The ground and excited electronic states correspond to different charge distributions. The minima of the potential energy curves correspond to atomic configurations with vanishing (net) dipole moment. The corresponding electric fields exert forces on ions and thereby (slightly) change their positions so that the potential energy curves corresponding to the ground and (first) excited are displaced both along the (horizontal) coordinate and the (vertical) energy axes.

The electronic transition is governed by the separation of energies between electrons in the ground state $|\psi_0\rangle$ and excited state $|\psi_\mu\rangle$, $\Delta E = E_\mu - E_0$. This separation is governed by the separation of ions, that is the bond lengths, it is much smaller when atoms (ions) are displaced from their equilibrium positions. Thus, for molecules, the electric susceptibility is characterized by the oscillator strength associated with the displacement of ions away from their equilibrium positions; that is, it is described by the linear response of molecular vibrations to an oscillatory forcing. Molecular vibrations can be described (or approximated) by harmonic oscillators; their dynamic susceptibility (linear response) is the same for classical and quantum oscillators are the same. It follows that the frequency-dependence is the same in the classical and quantum description $\chi \sim 1/(\omega_n^2 - \omega^2 - i\omega\gamma_n)$ for the oscillator labeled μ. In a classical description, the contribution of an oscillator is determined by the dipole moment $p = ed$ associated with the displacement d of atoms (ions), whereas in a quantum-mechanical description it is determined by the expectation value between the electron ground $|\psi_0\rangle$ and excited state $|\psi_\mu\rangle$

$$f_\mu \sim |\vec{p}|^2 \sim |\langle\psi_\mu|\vec{p}|\psi_0\rangle|^2. \tag{2.27}$$

The coefficients obey the sum rule

$$\sum_\mu f_\mu = Z. \tag{2.28}$$

This condition is quantum-mechanical in origin: it is derived from the condition that the simultaneous precise measurements of the position and momentum of a particle are impossible in quantum mechanics (Heisenberg uncertainty principle or equivalently the commutator relation between position and momentum operators).[1]

The overall molecular response is the sum of the contributions from each oscillator determined by the dynamic susceptibility that corresponds to replacing the (inverse) spring constant by a frequency-dependent term, $d \sim 1/k \rightarrow (1/m)\chi(\omega)$. We thus find

$$\chi(\omega) \sim \frac{ne^2}{m} \sum_\mu f_\mu \frac{1}{\omega_\mu^2 - \omega^2 - i\omega\gamma_\mu}. \tag{2.29}$$

In this expression, there is no quantity that has a genuine quantum-mechanical origin; this explains why (ordinary) frequency dispersion is generally described in terms of classical description of the motion of atoms (ions). A 'mixed' description—that is, quantum mechanics for electrons and classical mechanics for atoms—is familiar from the Born-Oppenheimer approximation.[2] In this sense,

[1]See, e.g., G. Baym, p. 281 ff.
[2]See, e.g., G. Baym, p. 471 ff.

a quantum-mechanical explanation of light absorption is possible but is not required. We note, however, this is not always correct: a quantum-mechanical treatment of atomic degrees of freedom is necessary for a description of IR spectra.

Light intensity is generally diminished as it travels through dielectric matter. The Beer-Lambert law states that the absorbance is proportional to the concentration of the absorbing substance. In a macroscopic description, light absorption can be described in terms of complex-valued wave vector

$$\mathsf{k} \longrightarrow \mathsf{k}' + i\,\mathsf{k}'' \tag{2.30}$$

with $\mathsf{k}' = \mathrm{Re}(\mathsf{k})$ and $\mathsf{k}'' = \mathrm{Im}(\mathsf{k})$. That is, a spatially varying electric field becomes exponentially decaying

$$e^{ikx} \longrightarrow e^{i\mathsf{k}'x}e^{-\mathsf{k}''x}. \tag{2.31}$$

Here, the imaginary part k'' is determined by the interaction of light with matter; in particular, light absorption involves a jump of an electron from its ground state to an excited state.

While we have ignored many (important) details of the response of dielectric matter to a time-dependent electric field, the important results are that the dielectric response is a scalar function so that the electric displacement and the electric field are always parallel to each other; details of the time-dependence or $\epsilon(t)$ or frequency-dependence $\epsilon(\omega)$ are determined by dynamic properties of molecules (or atoms). The real part $\mathrm{Re}\epsilon(\omega)$ determines the speed of propagation of electromagnetic waves in matter, while the imaginary part $\mathrm{Im}\epsilon(\omega)$ determines absorption.

2.6 Outlook

Since the electric susceptibility χ determines the (relative) electric permeability ϵ, Eq. (2.29) is the starting point for discussion of the time- (or frequency-) dependent response of dielectric matter in standard electrodynamics textbooks and is not repeated here in detail. The real part of the dynamic susceptibility determines the amplitude of the driven oscillator, while the imaginary part determines the decay of the amplitude. The Faltungs theorem for Fourier transforms yields for the electric displacement

$$\vec{D}(\omega) = \epsilon(\omega)\vec{E}(\omega). \tag{2.32}$$

where the dielectric susceptibility has both a real and an imaginary part

$$\epsilon(\omega) = \text{Re}\epsilon(\omega) + i\text{Im}\epsilon(\omega). \tag{2.33}$$

The real part determines the magnitude of the field $|\vec{D}|$ and the imaginary part determines dissipation of a time-dependent electric field incident on the dielectric matter. Since the imaginary part can be written as the difference between the dielectric susceptibility and its complex conjugate $2\cdot\text{Im}\epsilon = \epsilon - \epsilon^*$; we recover the expression for the energy dissipation of an electromagnetic wave, that is the absorption of light, briefly mentioned in Sect. 1.4.

We conclude that the molecular dipole moment induced by an impeding electromagnetic wave is independent of the polarization: it is the same for linear polarized light (along perpendicular axis $\hat{x}$ and $\hat{y}$ with $\vec{\mathsf{k}}\cdot\hat{x} = \vec{\mathsf{k}}\cdot\hat{y}$) and right- and left-circularly polarized light. It follows, in particular, that the polarization $\vec{P}$ and thus the electric displacement $\vec{D}$ is parallel to the electric field $\vec{D} \parallel \vec{E}$ so that the dielectric properties are described by a *scalar* electric susceptibility. The time-dependence does not upend the scalar relation; and we have for a fixed frequency ω

$$\vec{D}(\omega) = \epsilon(\omega)\vec{E}(\omega). \tag{2.34}$$

that is, in the time-domaine the electric displacement depends on the electric field at all previous times, cf. Eq. 1.21,

$$\vec{D}(t) = \int_{-\infty}^{t} \epsilon(t - t')\vec{E}(t')dt', \tag{2.35}$$

or alternatively,

$$\vec{D}(t) = \int_{-\infty}^{t} \epsilon(t')\vec{E}(t - t')dt'. \tag{2.36}$$

The key result is the scalar 'character' of the response of dielectric matter. This implies, in particular, that response is the same for all types of polarization (linear polarizations along perpendicular directions and right- and left-circular polarizations). It follows that the interaction of light with dielectric matter does not change the polarization: if linearly polarized light along the x-axis in air is incident on dielectric matter, the light remains polarized along the x-axis. That is, the description of frequency dispersion cannot be modified in some fashion to explain optical activity.

2.7 Annotated Bibliography

Dielectric matter: Properties of dielectric matter is discussed in introductory texts, e.g., J. Walker, *Halliday/Resnick/Walker—Fundamentals of Physics* 7th Ed. (J. Wiley & Sons, New York, 2005), Sect. 25-6, and R. P. Feynman, R. B. Leighton, and M. Sands, *Feynman Lectures on Physics* Vol. II (Addison-Wesley, Reading, MA, 1964), Chapter 10, and undergraduate texts, e.g., D. J. Griffiths, *Introduction to Electrodynamics* 3rd Ed. (Prentice Hall, Upper Saddle River, NJ, 1999), Chapter 4 and E. M. Purcell and D. J. Morin, *Electricity and Magnetism* 3rd Ed. (Cambridge University Press, New York, 2013), Chapter 10. Dielectric matter is discussed in detail in graduate texts J. D. Jackson, *Classical Electrodynamics* 3rd Ed. (J. Wiley & Sons, New York, 1999), Chapter 6 and A. Zangwill, *Modern Electrodynamics* (Cambridge University Press, New York, 2013), Chapter 6.

Frequency dispersion: Frequency dispersion, or just simply dispersion, is discussed in introductory texts, e.g., J. Walker, *Halliday/Resnick/Walker—Fundamentals of Physics* 7th Ed. (J. Wiley & Sons, New York, 2005), Sect. 33-8, and undergraduate texts, e.g., D. J. Griffiths, *Introduction to Electrodynamics* 3rd Ed. (Prentice Hall, Upper Saddle River, NJ, 1999), Sect. 9.4, and R. P. Feynman, R. B. Leighton, and M. Sands, *Feynman Lectures on Physics* Vol. II (Addison-Wesley, Reading, MA, 1964), Chapter 32. Dispersion is discussed in detail graduate level texts J. D. Jackson, *Classical Electrodynamics* 3rd Ed. (J. Wiley & Sons, New York, 1999), and A. Zangwill, *Modern Electrodynamics* (Cambridge University Press, New York, 2013).

Einstein coefficients: The basics of spontaneous and stimulated emission is discussed in Modern Physics texts such as R. Eisberg and R. Resnick, *Quantum Physics of Atoms, Molecules, Solids, Nuclei, and Particles* 2nd Ed. (J. Wiley & Sons, New York, 1985); Sect. 11-7.

3

Magneto-optic Effect

The phenomenon 'optical activity' describes the rotation of the polarization axis of linear polarized light in the plane perpendicular the wave vector $\vec{\mathsf{k}}$ in either counterclockwise or clockwise direction 'looking into the oncoming wave.' So, obviously, optical activity has 'something to do' with the handedness of the matter. In Sect. 3.1, we highlight the connection between optical activity and handedness by describing linear polarized light as a superposition of right- and left-circularly polarized light. We then discuss in Sect. 3.2 how an applied magnetic field is an *external* mechanism for rotating the axis of linear polarization. In Sect. 3.3, we use this insight to develop a macroscopic description of optical activity that provides clues into an *internal* mechanism for optical activity.

3.1 Phenomenological Description

Optical activity refers to the rotation of polarization axis of linearly polarized light in either right- or left-handed direction with respect to $\hat{\mathsf{k}}$. The handedness associated with optical activity suggests a connection with right- and left-circularly polarized light. We discussed in Sect. 1.4 that the superposition of right- and left-circularly polarized light yields linearly polarized light. That is, while the circularly-polarized light have components along both the x- and y-axes (assuming that the wave travels along the z-axis), the components along the y-axis, say, cancel out each other.

We now describe the rotation of the polarization axis when the cancellation of the right- and left-circularly polarized light along the y-axis is no longer 'perfect,' $\vec{E}(\vec{r},t) = \vec{E}_r(\vec{r},t) + \vec{E}_l(\vec{r},t) \neq E(\vec{r},t)\hat{x}$, that is, there is difference in the phase between the contributions to the electric field $\vec{E}_r$ and $\vec{E}_l$. In general, the phase has two distinct contributions determined by the wave vector and

DOI: 10.1201/9781003560944-3

the frequency $\exp(i\vec{\mathsf{k}}\cdot\vec{r})$ and $\exp(-i\omega t)$, respectively. A phase difference $\exp(i[\vec{\mathsf{k}}_r - \vec{\mathsf{k}}_l'] \cdot \vec{r})$ based on the difference between wave vectors would reflect a mechanism dominated by contribution as light travels between molecules. However, optical activity is based on the interaction between incident light on certain dielectric matter: that is, molecules of some dielectric matter have the 'ability' to rotate the electric field. We conclude that optical activity is described *phenomenologically* by a phase difference based on the difference between frequencies of right- and left-circularly polarized light $\exp(-i[\omega_r - \omega_l]t)$.

We now repeat the superposition of right- and left-circularly polarized for the cases $\omega_l > \omega_r$ and $\omega_r > \omega_l$; we see in Fig. 3.1 that the electric field rotates in *counterclockwise* and *clockwise* direction respectively. We write the rotating electric field vector in terms of time-dependent x- and y-components. We assume that the electric field is directed along the x-axis at the time $t = 0$ so that for $t > 0$, $\{E_{r,x} = E_0\cos(\omega_r t), E_{r,y} = -E_0\sin(\omega_r t)\}$, and $\{E_{l,x} = E_0\cos(\omega_l t), E_{l,y} = E_0\sin(\omega_l t)\}$. The superposition is then $\{E_x = E_{r,x} + E_{l,x}, E_y = E_{r,y} + E_{l,y}\}$. We use the trigonometric identities $\sin\alpha - \sin\beta = 2\cos([\alpha+\beta]/2)\sin([\alpha-\beta]/2)$ and $\cos\alpha + \cos\beta = 2\cos[\alpha+\beta]/2)\cos([\alpha-\beta]/2)$ and find

$$\begin{aligned} E_x(t) &= 2E_0\cos\left(\frac{[\omega_r+\omega_l]\,t}{2}\right)\cos\left(\frac{[\omega_r-\omega_l]\,t}{2}\right), \\ &\simeq 2E_0\cos(\omega\,t)\cos(\Delta\omega\,t)\,; \end{aligned} \tag{3.1}$$

$$\begin{aligned} E_y(t) &= 2E_0\sin\left(\frac{[\omega_r+\omega_l]\,t}{2}\right)\cos\left(\frac{[\omega_r-\omega_l]\,t}{2}\right) \\ &\simeq 2E_0\sin(\omega\,t)\cos(\Delta\omega\,t)\,, \end{aligned} \tag{3.2}$$

where we define the average frequency and the difference frequency

$$\omega = \frac{\omega_r+\omega_l}{2} \quad \text{and} \quad \Delta\omega = \frac{\omega_r-\omega_l}{2}. \tag{3.3}$$

We thus have the electric field:

$$\begin{aligned} \vec{E}(t) &= 2E_0\cos(\omega\,t)\cdot[\cos(\Delta\omega\,t)\,\hat{x} + \sin(\Delta\omega\,t)\,\hat{y}]\,, \\ &= E(t)\,\hat{\zeta}(t), \end{aligned} \tag{3.4}$$

where we define the amplitude

$$E(t) = 2E_0\cos(\omega\,t); \tag{3.5}$$

and the unit vector in the (x, y)-plane

$$\hat{\zeta}(t) = \cos(\Delta\omega\, t)\,\hat{x} + \sin(\Delta\omega\, t)\,\hat{y}. \tag{3.6}$$

That is, the magnitude of the electric field oscillates with frequency ω and the polarization rotates around the z-axis with frequency $\Delta\omega$.

Since the periods characterizing circularly polarized light $T_{r,l} = 2\pi/\omega_{r,l}$ are nearly identical $T_r \simeq T_l$ so that the period for the magnitude follows $T = 2\pi/\omega \simeq T_r, T_l$. The period associated with the rotating unit vector follows $\tau = 2\pi/\Delta\omega$. We find

$$\tau = 2\left(\frac{1}{T_r} - \frac{1}{T_l}\right)^{-1}. \tag{3.7}$$

We introduce the difference of periods $\Delta T = T_l - T_r$ and obtain $\tau = 2T^2/\Delta T$ so that

$$\frac{\tau}{T} = 2\frac{T}{\Delta T}. \tag{3.8}$$

Since the difference of the periods for right- and left-circularly polarized light is small $\Delta T << T$ so that

$$\frac{\tau}{T} >> 1, \tag{3.9}$$

which is in agreement with the expected separation of time scale between the oscillation of the electric field along the temporary polarization axis and the rotation of the that axis.

Thus, the rotation of the polarization axis can be attributed to a difference between the index of refraction of right- and left-circularly polarized light; the polarization axis rotates counter-clockwise (clockwise) for $n_l < n_r$ ($n_l > n_r$). The small difference between frequencies corresponds to a small difference of the index of refraction $\Delta n/n = (n_l - n_r)/n << 1$. We have $\Delta\omega = ck(1/n_r - 1/n_l)/2 = ck\,(\Delta n/2n^2)$ with $n \simeq n_r \simeq n_l$. We thus have for the ratios $\Delta\omega/\omega = \Delta n/n$. We assume that the sample has thickness d; we calculate the corresponding angle $\Delta\phi$ through which the polarization axis rotates. The time necessary for the wave to travel through the sample $\Delta t = d/v$ so that $\Delta\phi = (\Delta n/n)\omega \cdot (d/v) = (\Delta n/n)\mathsf{k}d$, or

$$\frac{\Delta\phi}{d} = \frac{\Delta n}{n}\mathsf{k}. \tag{3.10}$$

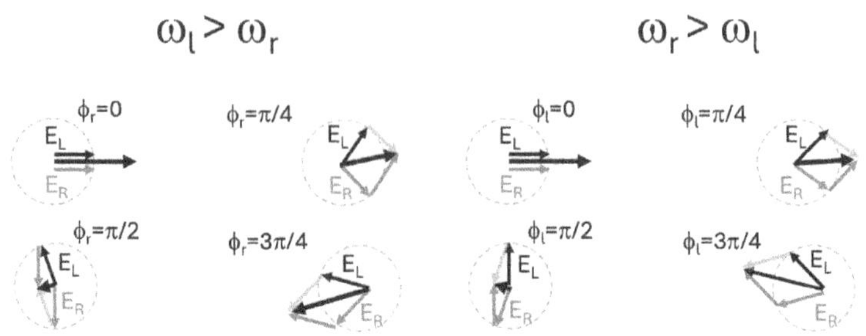

FIGURE 3.1
The superposition of the electric field vectors associated with right- (gray) and left- (black) circularly-polarized light $\vec{E} = \vec{E}_r + \vec{E}_l$. The electric field $\vec{E}$ rotates *counterclockwise* for $\omega_l > \omega_r$ and *clockwise* for $\omega_r > \omega_l$.

Thus, the frequency of optical activity enters via the explicit dependence on the wave vector and implicitly via the index of refraction.

Numerical values: Kauzmann *et al.*[1] write that 'butyl alcohol [...] ha[s] a molecular rotation of 10.3° in sodium D light at 20°C.' Here, the molecular rotation is reported when light travels through a sample with distance of one *deci*-meter (dm) so that it is assumed $d = 0.1\,\mathrm{m}$. Since $\Delta\phi = 10.3° \simeq 0.18\,\mathrm{rad}$, we have

$$\frac{\Delta\phi}{d} = \frac{0.18\,\mathrm{rad}}{0.1\,\mathrm{m}} = 1.8\,\mathrm{m}^{-1} \tag{3.11}$$

for butyl alcohol. The value for glucose is reported as 52.7° so that $\Delta\phi/d = 0.92\,\mathrm{m}^{-1}$. We thus see that the numerical value optical activity depends on the type of molecule.

The sodium D-line have wavelength $\lambda = 589.0\,\mathrm{nm}$ (589.6 nm) corresponding to a wave vector $\mathsf{k} = 1.07 \times 10^7\,\mathrm{m}^{-1}$. We thus find the estimate for butyl alcohol $\Delta n/n = \mathsf{k}^{-1} \cdot (\Delta\phi/d)$ so that

$$\frac{\Delta n}{n} = \left(1.07 \times 10^7\,\mathrm{m}^{-1}\right)^{-1} \cdot 0.18\,\mathrm{m}^{-1} \simeq 0.168 \times 10^{-6} \tag{3.12}$$

and for glucose $\Delta n/n = 0.863 \times 10^{-6}$. We assume $n \simeq 1$. Since $n \sim \sqrt{\epsilon}$, we have a small variation of the index of refraction corresponds to a variation of the *relative* dielectric constant from unity

$$\epsilon_{\mathrm{rel}} = 1 + 2\Delta n. \tag{3.13}$$

[1]W. J. Kauzmann, J. E. Walter, and H. Eyring, Chem. Rev. **26**, 339 (1940).

That is, molecules respond to the electric field of the incident light with a polarization that is exceedingly small: the variation of the index of refraction is of the order of *parts per million* (ppm). Such a variation is generally not observable; it is relevant in the context of optical activity since it produces a novel effect.

A calculation of exact values of optical activity requires a detailed quantum mechanical calculation and is outside the scope of this book. Instead we seek to understand how the 'special' character of the polarization and the small value of $\Delta n/n$ are closely connected to each other. That is, the mechanism that explains mechanism that allows a molecule to rotate the electric field of incident light also explains its small value.

3.2 Faraday Effect

We seek a mechanism that breaks this right-/left-handed symmetry of the response of the molecule. We distinguish between external and internal mechanisms. While optical activity is due to an internal mechanism, we discuss in this chapter an external mechanism as it provides us some insights into optical activity, as we discuss below. We recall that axial vectors, such as magnetic fields, define handedness, the obvious choice of an external mechanism is an applied static magnetic field. In the literature, this mechanism is referred to as magneto-optical effect.

Since $\vec{E} \times \vec{g} = -\vec{g} \times \vec{E}$ and $\vec{k} || \vec{g}$, we arrive at the polarization $P \sim \vec{k} \times \vec{E}$. Since g/ϵ_0 is dimensionless,

$$\vec{P} = -i\epsilon_0 l\vec{k} \times \vec{E}, \tag{3.14}$$

where we introduce the length scale l.

Since the charge q rotates around the center, the magnetic force $\vec{F} = q\vec{v} \times \vec{B}_{\text{ext}}$ is directed in radial direction; the force is directed *inward* (*outward*) for right- and left-circular polarization. That is, an external magnetic field $\vec{B} || \vec{\mathsf{k}}$ is compatible with an electric field rotating about direction of propagation, that is, with circularly-polarized light.

This is at variance with linearly-polarized light. To see this, we consider a polarization along the x-axis, so that the bond oscillates along the x-axis. The magnetic force is directed along the y-axis and thus perpendicular to the dipole moment of the molecule. Thus a constant magnetic field generates inertia for circularly-polarized but not for linearly-polarized light. It follows that the presence a magnetic field 'destroys' the equivalence of the pairs of polarization 'states' $(\hat{e}_x, \hat{e}_y)$ and $(\hat{e}_\mathrm{r}, \hat{e}_\mathrm{l})$. Light initially polarized along the x-axis will rotate and become polarized along the y-axis, whereas right-circularly polarized light will remain right-circularly polarized.

We first discuss *qualitatively* the effect of an external magnetic field on the propagation of circularly-polarized light. For right- (left-) circularly polarized light, the magnetic force is directed opposite (along) the electric force and thus magnetic field decreases (increases) the bond length $r \to r \mp \delta r$, and thus changing the moment of inertia of the rotating charged particle $I = mr^2 \to I \mp \delta I$ with $\delta I = 2mr\delta r$; here the negative (positive) sign applies for left- (right-) circularly polarized light. Since the (wave) speed is inversely proportional to the inertia, we expect that right-circularly polarized travels faster than left-circularly polarized light: that is, the frequency for right-circularly polarized light is greater than the frequency of left-circularly polarized light $\omega_r > \omega_l$.

The derivation is straightforward. We introduce the frequency

$$\omega_B = \frac{qB_\mathrm{ext}}{m} \tag{3.15}$$

(sometimes referred to as *cyclotron* frequency) and find

$$m\omega^2 r = m\omega_0^2 r - qE \pm m\omega_B\,\omega r, \qquad \left\{\begin{array}{cc} + & \text{right-circular} \\ - & \text{left-circular} \end{array}\right. \tag{3.16}$$

so that the bond is shortened or lengthened in the presence of an external magnetic field. We find

$$r = \frac{qE}{m(\omega_0^2 - \omega^2 \mp \omega_B\omega)} \qquad \left\{\begin{array}{cc} - & \text{right-circular} \\ + & \text{left-circular} \end{array}\right. . \tag{3.17}$$

Following the discussion in Sect. 1.5, we seek a description of the dynamic susceptibility in terms of the frequency ω of light. We simply complete the square and find

$$r = \frac{qE}{m([\omega_0^2 + \omega_B^2/4] - [\omega \pm \omega_B/2]^2)} \qquad \left\{\begin{array}{cc} + & \text{right-circular} \\ - & \text{left-circular} \end{array}\right. . \tag{3.18}$$

We find

$$\omega_{r,l} = \omega \pm \frac{\omega_B}{2} \tag{3.19}$$

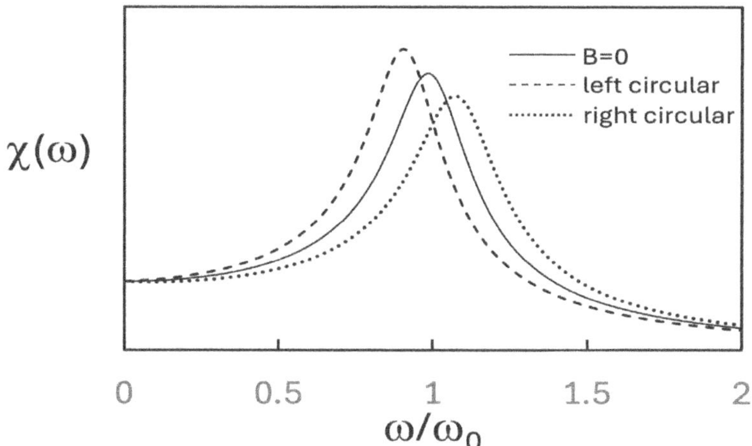

FIGURE 3.2
The amplitude of the dynamic susceptbility of a bond for zero magnetic field $\vec{B} = 0$ (solid) and in the presence of a magnetic field $\vec{B}$ parallel to the direction of propagation k for right- (dashed) and left- (dotted) circularly polarized light.

(for $\omega_L/\omega << 1$). In Fig. 3.2, we illustrate the frequency-dependence of the real part of the dynamic susceptibility for right- and left-circularly polarized light $\mathrm{Re}[(\omega^2 - \omega_{r,l}^2 - i\gamma\omega]^{-1}$ [for some non-zero-damping $\gamma \neq 0$], the discussion in Sect. 1.1. We recover the anticipated result: a constant magnetic field aligned with the direction of propagation speeds (slows down) right- (left-) circularly polarized light.

We write the wave speed $v_{r,l} = \omega_{r,l}/\mathsf{k}$ so that $\omega_{r,l}/\mathsf{k} = (\omega/\mathsf{k}) \cdot (1 \pm \omega_B/2\omega) = c\,(1 \pm \omega_B/2\omega)$. Since $v = c/n$, we find the index of refraction corresponding to right- and left-circularly polarized light

$$n_{r,l} = \left(1 \pm \frac{\omega_B}{2\omega}\right)^{-1} \simeq 1 \mp \frac{\omega_B}{2\omega}, \tag{3.20}$$

where we used $\omega_B < \omega$. In turn, the index of refraction is determined by the relative electric susceptibility $n = \sqrt{\epsilon_{\mathrm{rel}}}$ so that the relative dielectric constant is different for right- and left-handed polarized light

$$\epsilon_{r,l} = 1 \mp \frac{\omega_B}{\omega}. \tag{3.21}$$

so that

$$\frac{\Delta n}{n} = \frac{\omega_B}{\omega}. \tag{3.22}$$

Alternatively, we identify the external magnetic field as the 'forcing' for the rotation of the polarization axis and write

$$\frac{\Delta n}{n} = \frac{B_{\text{ext}}}{\mathcal{B}}, \tag{3.23}$$

where we introduced the 'characteristic' magnetic field

$$\mathcal{B} = \frac{m\omega}{q}. \tag{3.24}$$

We recall that for circular polarization light rotates the charge q with an angular speed equal to the frequency of the incident light. Here, $\mathcal{B}$ is the magnetic field such that the corresponding cyclotron frequency is equal to the frequency of light $\omega_B(\mathcal{B}) = \omega$. We emphasize that $\mathcal{B}$ is unrelated to the magnetic field associated with incident electromagnetic wave.

Numerical values: We seek a large 'effect,' that is a large value of $\Delta n/n$. Since the cyclotron frequency is proportional to the magnetic field $\omega_B \sim B$, we choose a large magnetic field $B \simeq 1\,\text{T}$, corresponding to the magnetic field of a (diagnostic) MRI (1 T) but larger than the Earth's magnetic field ($50\,\mu\text{T}$). We assume a large magnetic field $B_0 \simeq 1\,\text{T}$ which is similar to the magnetic field used in diagnostic MRI and is several orders of magnitude larger than the Earth's magnetic field $B_{\text{Earth}} \simeq 50\,\mu\text{T}$. We find the Larmor frequency for an elementary charge $q = e$ and the mass $m = \mathcal{M}$,

$$\omega_B \simeq \frac{1.6 \times 10^{-19}\,\text{C} \cdot 1\,\text{T}}{1.67 \times 10^{-27}\,\text{kg}} \simeq 10^8\,\text{s}^{-1}. \tag{3.25}$$

We use visible light $\lambda = 500\,\text{m}$ so that $\omega = 6 \times 10^{14}\,\text{s}^{-1}$ so that the ratio follows

$$\frac{\Delta n}{n} = \frac{10^8\,\text{s}^{-1}}{10^{14}\,\text{s}^{-1}} = 10^{-6}. \tag{3.26}$$

We calculate the characteristic magnetic field

$$\mathcal{B} = \frac{1.67 \times 10^{-27}\,\text{kg} \cdot 10^{14}\,\text{s}^{-1}}{1.6 \times 10^{-19}\,\text{C}} = 10^6\,\text{T}, \tag{3.27}$$

which is an enormous value. We recover the ratio

$$\frac{\Delta n}{n} = \frac{1\,\text{T}}{10^6\,\text{T}} = 10^{-6}. \tag{3.28}$$

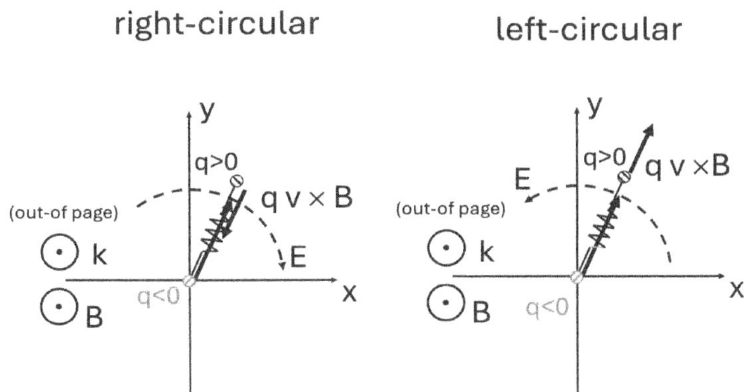

FIGURE 3.3
For right-circularly polarized light, the bond (harmonic oscillator) rotates clockwise and the magnetic field points out-of-the-page. The magnetic force acts toward the center. The external magnetic field must be flipped $\vec{B}_{\text{ext}} \to -\vec{B}_{\text{ext}}$ under inversion $\vec{r} \to -\vec{r}$.

Handedness: Handedness and inversion symmetry are closely related to each other; we have seen, in particular, that a right- (left-) circularly polarized wave transforms into a left- (right-) circularly polarized wave under inversion. In Figs. 3.3 and 3.4, we consider two distinct situations: (i) a right-circularly wave traveling along the $+z$-axis $\vec{\mathsf{k}} = \mathsf{k}\hat{z}$ (i.e., out of the page) ('original') and (ii) a right-circularly polarized wave traveling along the $-z$-axis $\vec{\mathsf{k}}' = -\mathsf{k}\hat{z}$ obtained by spatial inversion of a left-circularly polarized wave traveling along $+z$-axis $\vec{\mathsf{k}} = \mathsf{k}\hat{z}$ ('inversion'). We see that the magnetic field must be flipped $\vec{B}'_{\text{ext}} = -\vec{B}_{\text{ext}}$ for the magnetic force being directed toward the center both in the 'original' and 'inversion' version. We leave it to the reader to check that we reach the same conclusion by considering a left-circularly polarized wave. We note that the flip of the magnetic field is consistent with the inversion symmetry of susceptibilities, cf. Sect. 1.1.

3.3 Continuous Description

We have seen in the preceding section that the presence of a static *magnetic* field directed along the direction of the wave vector $\vec{\mathsf{k}}$

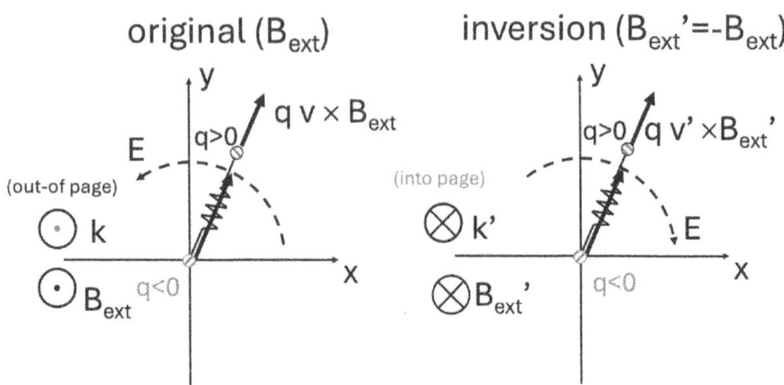

FIGURE 3.4
For left-circularly polarized light, the bond (harmonic oscillator) rotates counterclockwise and the magnetic field points out-of-the-page. The magnetic force acts away from the center. The external magnetic field must be flipped $\vec{B}_{\text{ext}} \to -\vec{B}_{\text{ext}}$ under inversion $\vec{r} \to -\vec{r}$.

rotates the electric field. In this section, we seek a macroscopic description of this magneto-optic *rotational power*. The complex notation of the electric field (while keeping in mind, of course, that only the real part corresponds to the observed electric field) since a rotation in counterclockwise (in clockwise) direction corresponds to a multiplication with the imaginary unit i ($-i$).

We start from a description of the electric displacement $\vec{D}$ in terms of the electric polarization $\vec{P}$ so that $\vec{D} = \epsilon_0\vec{E} + \vec{P}$. We introduce the 'gyration' vector

$$\vec{g} = g\hat{z}; \tag{3.29}$$

where the magnitude is proportional to the external magnetic field, $g \sim B$. We note that the vector product $\vec{E} \times \vec{g}$ points in *clockwise* direction away from the direction of the electric field. Since the polarization describes a rotation, we write

$$\vec{P} = i(\vec{E} \times \vec{g}). \tag{3.30}$$

Note that the gyration vector g and the constant ϵ_0 have the same unit $[g] = [\epsilon_0] = \mathrm{C}^2/(\mathrm{N\,m}^2)$.

We now show that this expression for the polarization yields the desired frequency dependence We assume a wave-like solution along the z-axis $\vec{\mathsf{k}} = \mathsf{k}\hat{z}$, $\vec{E}(\vec{r},t) = \vec{E}\exp(i[\mathsf{k}z - \omega t])$. We use the substitutions for the divergence and curl, $\nabla\cdot \to i\vec{\mathsf{k}}\cdot$ and $\nabla\times \to i\vec{\mathsf{k}}\times$ for spatial derivatives and the substitution $\partial/\partial t \to -i\omega$ for time derivatives. Faraday's law and the displacement current now follow

$$\vec{\mathsf{k}} \times \vec{E} - \omega\vec{B} = 0, \tag{3.31}$$

$$\frac{1}{\mu_0}\vec{\mathsf{k}} \times \vec{B} + \omega\vec{D} = 0. \tag{3.32}$$

We have $\vec{\mathsf{k}} \times (\vec{\mathsf{k}} \times \vec{E}) = \vec{\mathsf{k}}(\vec{\mathsf{k}} \cdot \vec{E}) - \mathsf{k}^2\vec{E} = \omega\vec{\mathsf{k}} \times \vec{B} = -\mu_0\omega^2\vec{D}$ (and using $\vec{\mathsf{k}} \cdot \vec{E} = 0$ for a transverse wave),

$$\mathsf{k}^2\vec{E} - \mu_0\omega^2\vec{D} = 0. \tag{3.33}$$

We insert $\vec{D} = \epsilon_0\vec{E} + i[\vec{E} \times \vec{g}]$ and find using $\mu_0 = (c^2\epsilon_0)^{-1}$,

$$\left(\mathsf{k}^2 - \frac{\omega^2}{c^2}\right)\vec{E} + i\left(\frac{\omega}{c}\right)^2\left[\vec{E} \times \left(\frac{\vec{g}}{\epsilon_0}\right)\right] = 0, \tag{3.34}$$

or, after multiplying by c^2/ω^2,

$$\left(\frac{\mathsf{k}^2c^2}{\omega^2} - 1\right)\vec{E} + i\left[\vec{E} \times \left(\frac{\vec{g}}{\epsilon_0}\right)\right] = 0. \tag{3.35}$$

We use the wave speed $v = c/n = \omega/\mathsf{k}$, where n is the index of refraction. We adopt the 'language' of dielectric matter and define the relative dielectric susceptibility $\epsilon_{\text{rel}} = n^2$ and find

$$(\epsilon_{\text{rel}} - 1)\vec{E} + i\left[\vec{E} \times \left(\frac{\vec{g}}{\epsilon_0}\right)\right] = 0. \tag{3.36}$$

We find in component form

$$(\epsilon_{\text{rel}} - 1)E_x - i\left(\frac{g}{\epsilon_0}\right)E_y = 0, \tag{3.37}$$

$$+i\left(\frac{g}{\epsilon_0}\right)E_x + (\epsilon_{\text{rel}} - 1)E_y = 0. \tag{3.38}$$

Eqs. (3.30) and (3.31) have the form of an eigenvalue equation

$$\begin{pmatrix} 1 & ig/\epsilon_0 \\ -ig/\epsilon_0 & 1 \end{pmatrix} \cdot \begin{pmatrix} E_x \\ E_y \end{pmatrix} = \epsilon_{\text{rel}}\begin{pmatrix} E_x \\ E_y \end{pmatrix}. \tag{3.39}$$

The 2×2 matrix $\mathbf{D}$ on the RHS is the generalization of the dielectric function. We have in cartesian (x, y)-coordinates (i.e., in the representation with respect to $[\hat{e}_x, \hat{e}_y]$ basis):

$$\epsilon = \begin{pmatrix} 1 & ig/\epsilon_0 \\ -ig/\epsilon_0 & 1 \end{pmatrix}. \tag{3.40}$$

We see by inspection that ϵ is a Hermitian matrix, cf. Sect. 1.3. The eigenvalues are real and the eigenvectors are orthogonal; this property is independent of the choice of basis vectors.

It is instructive to first consider the trivial case $g = 0$. The characteristic equation follows

$$[1 - \epsilon_{\text{rel}}]^2 = 0 \tag{3.41}$$

so that the frequency is given by the familiar expression $\epsilon_{\text{rel}} = 1$. There are two sets of eigenvectors

$$\left\{ \hat{e}_x = \begin{pmatrix} 1 \\ 0 \end{pmatrix}, \hat{e}_y = \begin{pmatrix} 0 \\ 1 \end{pmatrix} \right\}, \tag{3.42}$$

which correspond to linear polarized light along the x- and y-axis respectively, and

$$\left\{ \hat{e}_r = \frac{1}{\sqrt{2}} \begin{pmatrix} 1 \\ -i \end{pmatrix}, \hat{e}_l = \frac{1}{\sqrt{2}} \begin{pmatrix} 1 \\ +i \end{pmatrix} \right\}. \tag{3.43}$$

The two different sets of eigenvectors are consistent with the description of linear polarized light in terms of the superposition of right- and left-circularly polarized light (and vice versa).

For $0 < g/\epsilon_0 << 1$, the set of eigenvectors $\{\hat{e}_x, \hat{e}_y\}$ are no longer eigenvectors of the dielectric matrix ϵ. The characteristic equation reads

$$[1 - \epsilon_{\text{rel}}]^2 - \left(\frac{g}{\epsilon_0}\right)^2 = 0 \tag{3.44}$$

so that

$$\epsilon_{\text{rel}} = \sqrt{1 \pm \frac{g}{\epsilon_0}}. \tag{3.45}$$

We find the corresponding eigenvectors correspond to right- and left-polarized light:

$$\epsilon_r = 1 - \frac{g}{2\epsilon_0} \quad \hat{e}_r = \frac{1}{\sqrt{2}} \begin{pmatrix} 1 \\ -i \end{pmatrix} \qquad \text{right circular,} \tag{3.46}$$

$$\epsilon_l = 1 + \frac{g}{2\epsilon_0} \quad \hat{e}_l = \frac{1}{\sqrt{2}} \begin{pmatrix} 1 \\ i \end{pmatrix} \qquad \text{left circular.} \tag{3.47}$$

We obtain the indices of refraction for right- and left-circular polarized light

$$n_{r,l} \simeq 1 \mp \frac{g}{2\epsilon_0}. \tag{3.48}$$

That is, for $g \neq 0$, only right- and left-circularly polarized light are eigenstates of the electric susceptibility matrix, whereas linearly polarized light along the x- and y-axis are no longer eigenstates. It follows that light polarized along the x-axis, say, as it enters dielectric matter can become polarized along the y-axis if the sample is sufficiently thick; the linear polarization rotates about the direction of propagation.

For the magneto-optic effect, we compare Eqs. (3.18) and (3.39) and find $g/\epsilon_0 = \omega_B/\omega$. We adopt that the external magnetic field is the 'force,' $g/\epsilon_0 = B_{\text{ext}}/\mathcal{B}$ with the characteristic field proportional to the atomic mass $\mathcal{B} \sim m$. We thus expect that optical activity can be described by a complex-valued off-diagonal term

$$\frac{g}{\epsilon_0} \sim \frac{\text{force}}{\text{inertia}}, \tag{3.49}$$

where 'force' and 'inertia' depend on the physical mechanism underlying optical rotation. We expect that the rotation is small independent of the 'nature' of the forcing.

$$\frac{g}{\epsilon_0} \simeq 10^{-6}. \tag{3.50}$$

For one rotation of the electric field for circular polarized light corresponds to a change of the period (expressed as 24 hours) by less than 1 second.

3.4 Insights for Optical Activity

Optical activity and magneto-optical effect are 'similar' in this sense and rotate the axis of linearly polarized light. While the corresponding mechanisms are entirely different (with optical activity based on an *internal* mechanism and the magneto-optic effect based on an *external* mechanism), the mechanisms also share some similarity. We note $\vec{E} \times \vec{g} = -\vec{g} \times \vec{E}$ and $\vec{g}||\vec{k}$, we have the polarization $\vec{P} \sim i\vec{k} \times \vec{E}$. Since the ratio g/ϵ_0 is dimensionless, we arrive at the expression

$$\vec{P} = -i\epsilon_0 l \vec{\mathsf{k}} \times \vec{E}, \tag{3.51}$$

where l is a length scale.

Eq. (3.51) implies a Fourier description of the time and spatial dependence (in which case the multiplication with the wavevector corresponds to taking the derivative $-i\mathsf{k} \leftrightarrow [\partial/\partial x]$ (where we choose x for the coordinate). We conclude that optical activity relates the spatial variation of the dielectric matter. That is, the electric properties of a molecule cannot be described as a 'point'-dipole; rather the structure of molecules enters the description. While all molecules have some spatial structure (i.e., some specific geometry in terms of bond lengths and angles), most molecules are *not* optically active. We conclude that molecules must have some unique configurational properties to generate a polarization perpendicular to the electric field of the incident light.

We now turn to the 'strength' of the rotatory power. We start from the estimate $g/\epsilon_0 \times 10^{-6}$. Since visible light ($\lambda = 500\,\mathrm{nm}$) corresponds to the wave vector $\mathsf{k} \simeq 10^7\,\mathrm{m}^{-1}$, we find the length scale characterizing optical rotatory power

$$l \simeq \frac{10^{-6}}{10^7\,\mathrm{m}^{-1}} = 10^{-13}\,\mathrm{m}. \tag{3.52}$$

That is, we recover the length scale associated with the displacements of atoms (ions) that 'facilitate' electronic transitions associated with a molecular dipole moment, cf. Sect. 2.5.

An induced dipole moment of molecules is based on a cooperative response of atoms (ions) and valence electrons. The displacement of atoms (ions) is characterized by the length scale l. The coupling between atomic and electronic degrees of freedom 'facilitates' an electronic transition from the ground to an excited state. These transitions are associated with displacements of the electrons of the order $L >> l$. While the molecular dipole moments originate from atomic displacements, electronic transitions act as 'boosters' of the characteristic length scale associated with the dipole moment $l \longrightarrow \mathcal{L}$.

While optical rotatory power similarly relates to electronic transformations driven by atomic displacements, the characteristic length scale l can be understood by the absence of the booster associated with an electronic transition from the ground to an excited states. This is confusing since the absence of an electronic transition would suggest that the electron remains in the ground state and does not support any electronic transformation.

Evidently, optical activity involves 'some kind' of electric transformation; we conclude that the electric transformation mirrors the displacements of atoms (ions). The central result of the subsequent chapters is that dynamics of ions in optically active molecules allows for a 'virtual' electronic transformations that is based on the quantum-mechanical nature of electrons that has no correspondence in the framework of classical mechanics. This explains the phrase 'optical activity has quantum-mechanical origin.'

We gain a further clue in the underlying mechanism by examining the character of the vectors involved in optical rotation. The dipole moment is a polar vector, whereas a rotation is described by an axial vector. In this light, the existence of a magneto-optic mechanism of optical rotation is not surprising because the magnetic field is an axial vector. That is, handedness is imposed by an external field. In contrast, in optically active molecules, handedness is generated by the dynamics of atoms (ions).

3.5 Annotated Bibliography

Magneto-optic effect: The magneto-optic effect is discussed in L. D. Landau and E. M. Lifshitz, *Electrodynamics of Continuous Media* Vol. VIII (Pergamon, Oxford, 1980); Sect. 101. A brief discussion is found in E. Hecht, *Optics* (Addison-Wesley, Reading, MA, 1987); Sect. 8.11.2. An older, but still valuable, discussion is found in P. Drude, *Theory of Optics* (Longman, New York, 1907); Ch. VII.

Quantum mechanics: Virtual transitions are central to many quantum-mechanical behavior and is at the heat of 'quantum weirdness.' See, e.g., P. Ball, *Beyond Weird* (University of Chicago Press, Chicago, 2018), and D. F. Styer, *The Strange World of Quantum Mechanics* (Cambridge University Press, Cambridge, 2000).

4

Spatial Diffusion

We have seen in the preceding chapter that optical activity can be described phenomenologically either in terms of a vector product with the wave vector $\vec{\mathsf{k}} \times \vec{E}$ or in terms of a curl $\nabla \times \vec{E}$. We start from a macroscopic description of dielectric matter by incorporating not only a time-dependent (corresponding to frequency dispersion), but also a space-dependent response. This explains why optical activity is also referred to as *spatial dispersion.*

This is a unique property, we expect that it can be explained in the same general framework as the one used for (ordinary) frequency dispersion of molecules. That is, we expect that optical activity is 'driven' by displacement of atoms (ions) away from their equilibrium positions.

We discussed in Chapter 2 that the displacement of atoms (ions) is characterized by the length scale $l \sim (m_e/\mathcal{M})\mathcal{L} \simeq 10^{-13}\,\mathrm{m}$. Thus for visible light ($\mathsf{k} \sim 10^7\,\mathrm{m}^{-1}$), the product $\mathsf{k}l$ is small, $\mathsf{k}\,l \simeq 10^7\,\mathrm{m}^{-1} \cdot 10^{-13}\,\mathrm{m} = 10^{-6} << 1$. That is, quadratic (and higher) terms $(\mathsf{k}l)^2 \sim 10^{-12}$ can be ignored; we conclude that optical activity is explained in part of *linear* optics.[1] We note that many electron effects are irrelevant so that the one-valence electron approximation remains sufficient.

Optical activity requires a 'special type' of displacement of atoms (ions), which, in turn, can only be supported in molecules *without* inversion symmetry. That is, only *chiral* molecules rotate the polarization axis. The focus in this chapter is, therefore, the molecular structure, that is, the locations of atoms (ions) in molecules. We gain insights how the molecular structure supports the dynamics of atoms (ions) that supports the quantum-mechanical nature of the electronic transition underlying the rotation of the optical axis.

[1] A typical nonlinear effect is frequency mixing (second harmonic generation); see, e.g., N. Bloembergen, *Nonlinear Optics* 4th Ed. (World Scientific, Singapore, 1996).

DOI: 10.1201/9781003560944-4

4.1 Nonlocal Electric Fields

We assume a monochromatic wave $\vec{E}(\vec{r},t) = \vec{E}_0 \exp[i(\vec{\mathsf{k}} \cdot \vec{r} - \omega t)]$. The interaction of light with the dielectric medium implies that electric displacement at time t and the position $\vec{r}$ are determined by the electric field at all previous times $t' < t$ and at all locations $\vec{r}'$; we thus have for the l component

$$D_l(\vec{r},t) = E_l(\vec{r},t) + \sum_{m=x,y,z} \int d\vec{r}' \int_{-\infty}^{t} f_{lm}(\vec{r},\vec{r}';t-t')E_m(\vec{r}',t'). \tag{4.1}$$

Since the average electronic properties are determined by the atoms (ions), the kernel $f_{lm}(\vec{r}-\vec{r}';t-t')$ depends on the molecular structure characterized by the positions of atoms (ions), $\{\vec{r}_j\}$,

$$f_{lm}(\vec{r};t) = f_{lm}(\vec{r};t|\{\vec{r}_j\}). \tag{4.2}$$

We assume spatial homogeneity[2] then the kernel depends only on the displacement $\vec{\rho} = \vec{r} - \vec{r}'$:

$$f_{lm}(\vec{r},\vec{r}';t-t') = f_{lm}(\vec{r}-\vec{r}';t-t'). \tag{4.3}$$

The RHS of Eq. (4.1) then has the mathematical form of a faltung (convolution); a Fourier transform with respect to both time and space yields

$$D_l(\vec{\mathsf{k}},\omega) = \sum_m \epsilon_{lm}(\vec{\mathsf{k}},\omega)E_m(\vec{\mathsf{k}},\omega), \tag{4.4}$$

where the electric permeability is given by

$$\epsilon_{lm}(\vec{\mathsf{k}},\omega) = \delta_{lm} + \int_{-\infty}^{\infty} ds \int d\vec{\mathsf{r}} f_{lm}(\vec{\rho},s)e^{-i(\vec{\mathsf{k}}\cdot\vec{\mathsf{r}}-\omega s)}. \tag{4.5}$$

That is, the frequency and wave vector-dependence of the electric permeability capture the non-locality of the dielectric response with respect to time and space, respectively.

We assume that the polarization along the y-direction (P_y) produced by an electric field along the x-direction (E_x) is identical with the polarization along the x-direction (P_x) produced by an electric field along the y-direction (E_y); that is, the matrix (tensor)

[2]This is equivalent with the statement in introductory physics that the choice of origin of the coordinate system can be chosen arbitrarily.

f_{lm} is symmetric $f_{lm} = f_{ml}$. We have the relation (assuming that there is no magnetic field $\vec{B} = 0$)

$$\epsilon_{lm}(-\vec{\mathsf{k}}, -\omega) = \epsilon^*_{ml}(\vec{\mathsf{k}}, \omega), \tag{4.6}$$

or in terms of real (ϵ') and imaginary parts (ϵ''), so that $\epsilon = \epsilon' + i\epsilon''$ with $\epsilon'_{lm}(-\vec{\mathsf{k}}, -\omega,) = \epsilon'_{ml}(-\vec{\mathsf{k}}, -\omega)$ and $\epsilon''_{lm}(-\vec{\mathsf{k}}, -\omega) = -\epsilon''_{ml}(-\vec{\mathsf{k}}, -\omega)$. Time-reversal invariance $f(-t) = f(t)$ yields

$$\epsilon_{lm}(\vec{\mathsf{k}}, \omega) = \epsilon_{ml}(-\vec{\mathsf{k}}, \omega). \tag{4.7}$$

Dielectric properties are determined by the interaction of electromagnetic waves with atoms; that is, the kernel is zero if the position vectors $\vec{r}$ and $\vec{r'}$ are separated by more than the size of a single molecule (or atom): that is, $f_{lm}(\vec{r}, \vec{r'}; \omega) \simeq 0$ for $|\vec{r} - \vec{r'}| > \mathcal{L}$ (where $\mathcal{L}$ is the length scale from Sect. 1.5). For visible light, the wavelength $\lambda \simeq 500\,\text{nm}$ is much longer than the characteristic length $\lambda >> \mathcal{L}$ and the difference between $\vec{r}$ and $\vec{r'}$ becomes irrelevant. Thus, molecules can be treated as 'point'-scatters, and can be described by a (harmonic) oscillator with frequency ω and damping constant γ but without spatial extent. We conclude that frequency dispersion corresponds to the long-wavelength limit $\lambda \to \infty$ or $\mathsf{k} \to 0$: The spatial dependence of the kernel is simplified by a 'delta' function, $f_{lm}(\vec{\rho}, s) \simeq \delta(\vec{\rho}, s)$ so that

$$f_{lm}(\vec{\mathsf{k}}; \omega) \longrightarrow f(\omega) \quad \text{for} \quad \mathsf{k}\mathcal{L} \to 0 \tag{4.8}$$

so that for the dielectric permeability

$$\epsilon(\vec{\mathsf{k}}, \omega) = \epsilon(\omega), \tag{4.9}$$

as we discussed in Sect. 2.

Thus, a polarization proportional to the wave vector $P \sim \mathsf{k}$, cf. Eq. (3.51), originates from a finite size of molecules. We seek the leading correction for finite wavevector $|\vec{\mathsf{k}}| \neq 0$). As we will see, the corrections depend on the symmetry of the molecule. We start from Eq. (4.5) and use Taylor series expansion for the dependence on wave vector k,

$$\exp(-i\vec{\mathsf{k}} \cdot \vec{\mathsf{r}}) = 1 - i\sum_{n} \mathsf{k}_n \mathsf{r}_n - \frac{1}{2}\sum_{n,n'} \mathsf{k}_n \mathsf{k}_{n'} \mathsf{r}_n \mathsf{r}_{n'} + \cdots . \tag{4.10}$$

and find the correction in second order of the wave vector k,

$$\begin{aligned}\chi(\omega) &= \chi^{(0)}(\omega) + \chi^{(1)}(\vec{\mathsf{k}},\omega) + \chi^{(2)}(\vec{\mathsf{k}},\omega) \\ &= \chi^{(0)}(\omega) - i\sum_{n} V_n(\omega)\mathsf{k}_n, -\frac{1}{2}\sum_{n,n'} W_{nn'}(\omega)\mathsf{k}_n\mathsf{k}_{n'}\end{aligned} \qquad (4.11)$$

where introduced the vector (or first-rank tensor) V_n

$$V_n(\omega) = \frac{1}{\epsilon_0}\int_{-\infty}^{\infty} dse^{i\omega s}\int d\vec{\mathsf{r}}\,\mathsf{r}_n f(\vec{\mathsf{r}};s), \qquad (4.12)$$

and the second-rank tensor $W_{n,n'}$,[3]

$$W_{nn'}(\omega) = \frac{1}{\epsilon_0}\int_{-\infty}^{\infty} dse^{i\omega s}\int d\vec{\mathsf{r}}\,\mathsf{r}_n\mathsf{r}_{n'} f(\vec{\mathsf{r}};s). \qquad (4.13)$$

We note that these tensors have the physical significance of appropriately averaged locations of charges, $V_n = \langle \mathsf{r}_n \rangle$ and $W_{nn'} = \langle \mathsf{r}_n\mathsf{r}_{n'} \rangle$.

Molecules with inversion symmetry: If a molecule has inversion symmetry, the molecule consists of pairs of charges $q_i = q_{i'}$ at $\vec{r}_{i'} = -\vec{r}_i$. It follows that the kernel $f(\vec{\mathsf{r}},s)$ is identical in the original and inverted coordinate system,

$$f(\vec{\mathsf{r}};s) = f(-\vec{\mathsf{r}};s). \qquad (4.14)$$

We thus have

$$\int d\vec{\mathsf{r}}\,\mathsf{r}_n f(\vec{\mathsf{r}};s) = \int d\vec{\mathsf{r}}(-\mathsf{r}_n) f(-\vec{\mathsf{r}};s) = -\int d\vec{\mathsf{r}}\,\mathsf{r}_n f(\vec{\mathsf{r}};s) \qquad (4.15)$$

that is, we arrive at the equation $V_n = -V_n$, which implies that the vector describing the response of the molecules vanishes[4]

$$V_n = 0. \qquad (4.16)$$

We follow the same reasoning to show that second-rank tensor does not vanish $W_{nn'} \neq 0$. That is, the leading correction to frequency dispersion is second-order the small parameter $\mathsf{k}\mathcal{L}$,

[3]In the general case, the first- and the second-rank tensors V_n and $W_{nn'}$ correspond to a third- and fourth-rank tensor V_{lmn} and $W_{lmnn'}$, respectively.

[4]This result is reminiscent of electrons in metals. Conduction electrons are described by plane wave (so-called Bloch waves).

$[\chi(\vec{\mathsf{k}},\omega) - \chi^{(0)}(\omega)] \sim (\mathsf{k}l)^2$, which explains why spatial diffusion is generally ignored.

Our argument is based on the assumption that the (point-) charges (i.e., atoms [ions] and valence electrons) are at their respective reference positions. We assume that the incident wave travels along the z-axis so that $\mathsf{k}_x = \mathsf{k}_y = 0$ and $\mathsf{k}_z = \mathsf{k}$ so that $\chi(\omega) = \chi^{(0)}(\omega) - iV(\omega)\mathsf{k}$. The z-component of the vector V follows (we simplify the notation and drop the subscript z, $V_z^{(1)} = V$),

$$V(\omega) = -\frac{1}{\epsilon_0}\int_{-\infty}^{\infty} ds e^{i\omega s} \int d\vec{\rho}\, z \sum_j \left[\nabla f_j(\vec{\mathsf{r}} - \vec{\mathsf{r}}_i^{(0)}) \cdot \delta\vec{\mathsf{r}}_j(s)\right], \quad (4.17)$$

where ∇_i is the gradient with respect to $\vec{\mathsf{r}}_j$. We take the average over all coordinates: only the z-component survives

$$V = -\epsilon_0^{-1} \sum_j \int_{-\infty}^{\infty} ds e^{i\omega s} \int dz \left\langle z(\partial f_j/\partial z)\right\rangle \cdot \left\langle \delta z_j(s)\right\rangle .$$

The term $z\partial f_j/\partial z$ involves displacements of charges away from their respective references configurations. The corresponding change in bond is associated with changes in the electronic properties that include 'jumps' to excited states and thus transfer of energy from the incident electromagnetic wave to the dielectric matter. That is, coupling between vibrational (or atomic) and electronic degrees of freedom leads to dissipation (damping); the vector V defines a length scale $V \sim l$. We arrive at a polarization

$$P = -i\epsilon_0\, \mathsf{k} l E; \quad (4.18)$$

we note that the multiplication by the wavevector corresponds to a derivate with respect to a position $-i\mathsf{k} \leftrightarrow \partial/\partial r$ so that $-i\mathsf{k}E \leftrightarrow \partial E/\partial r$. That is, the polarization is proportional to the divergence of the electric field $P = \epsilon_0 l \nabla \cdot \vec{E}$. We use Poisson's law and introduce the induced charge density $\nabla \cdot \vec{E} = \rho_{\text{ind}}/\epsilon_0$ so that the polarization is given by $P = l\rho_{\text{ind}}$. Since charge density is equal to the induced charge per volume $\rho_{\text{ind}} = Q_{\text{ind}}/\text{volume}$, the above expression is consistent with the polarization as a dipole moment $p_{\text{ind}} = Q_{\text{ind}} l$ per volume, $P = p_{\text{ind}}/\text{volume}$.

The electric permeability is a scalar with an imaginary part proportional to the wave vector $\epsilon = \epsilon_0(1 - il\mathsf{k})$. We introduce the parameter $\zeta = l\mathsf{k}$ analogous to quantity g/ϵ_0 for the magneto-optic effect.

The electric displacement follows

$$D(\mathsf{k},\omega) = \epsilon_0(1+i\zeta)E(\mathsf{k},\omega). \tag{4.19}$$

Inserted into Eq. (3.32) we find $\mathsf{k}^2E - (\omega^2/c^2)(1+i\zeta)E = 0$. We introduce the wave vector corresponding the frequency ω in free space $\mathsf{k}_0 = \omega/c$

$$\left[\left(\frac{\mathsf{k}}{\mathsf{k}_0}\right)^2 - 1 - i\zeta\left(\frac{\mathsf{k}}{\mathsf{k}_0}\right)\right]E = 0. \tag{4.20}$$

We arrive at a quadratic equation for k/k_0:

$$\frac{\mathsf{k}}{\mathsf{k}_0} = +\frac{i\zeta}{2} \pm \sqrt{1-\zeta^2} \simeq \pm 1 - \frac{i\zeta}{2}. \tag{4.21}$$

Thus the wave vector has real and imaginary parts

$$\vec{\mathsf{k}} = \mathsf{k}_0(1+i\zeta) = \vec{\mathsf{k}}' + i\vec{\mathsf{k}}''. \tag{4.22}$$

The imaginary part describes exponential decay

$$e^{i\vec{\mathsf{k}}\cdot\vec{r}} = e^{-\vec{\mathsf{k}}''\cdot\vec{r}}e^{i\vec{\mathsf{k}}\cdot\vec{r}}. \tag{4.23}$$

We conclude that for symmetric molecules, a dielectric constant linear in the wave vector describes dissipation; at a molecular level, dissipation involves the creation of an electric dipole moment by the displacement of atoms (ions) away from their respective equilibrium positions.

We assume that the dielectric response is described by a single length scale; that is, the same length scale characterizes dissipation and optical rotation. We thus have $\zeta \sim 10^{-6}$ so that the imaginary component has the value $\mathsf{k}'' \simeq 10^{-6}\cdot 10^{7}\,\mathrm{m}^{-1} = 10\,\mathrm{m}^{-1}$. We calculate the decay of the electric field: we use the sample thickness used in the definition of specific optical rotation $d = 10\,\mathrm{dm} = 0.1\,\mathrm{m}$ and obtain $\mathsf{k}''d = 10\,\mathrm{m}^{-1}\cdot 0.1\,\mathrm{m} = 1$ so that $e^{-\mathsf{k}''d} = e^{-1} = 0.36$. This is consistent with the observation that optically active molecules are generally also good light absorbers.

Molecules without inversion symmetry: Some molecules do not possess inversion symmetry; an example is alanine shown in Fig. 4.1. In particular, many large molecules consist of a small pocket, a so-called *chromophore*; that is, light sensitive and is asymmetric with respect to inversion. The rest of the molecule is however symmetric with respect to inversion. We write the kernel as a sum of a symmetric and an antisymmetric term

$$\begin{aligned} f(\vec{\mathrm{r}};s) &= \frac{1}{2}\left[f(\vec{\mathrm{r}};s)+f(-\vec{\mathrm{r}};s)\right]+\frac{1}{2}\left[f(\vec{\mathrm{r}};s)-f(-\vec{\mathrm{r}};s)\right] \\ &= f^{(s)}(\vec{\mathrm{r}};s)+f^{(a)}(\vec{\mathrm{r}};s), \end{aligned} \tag{4.24}$$

where

$$f^{(s)}(\vec{\mathrm{r}};s)=f^{(s)}(-\vec{\mathrm{r}};s) \quad f^{(a)}(\vec{\mathrm{r}};s)=-f^{(a)}(-\vec{\mathrm{r}};s). \tag{4.25}$$

The superposition principle implies that the two terms correspond to two distinct contributions of the polarization

$$\vec{P}(\vec{r},t)=\vec{P}^{(s)}(\vec{r},t)+\vec{P}^{(a)}(\vec{r},t). \tag{4.26}$$

As shown above, the symmetric contribution describes absorption; we ignore absorption and set $\vec{P}^{(s)}(\vec{r},t)=0$. We conclude that the unique properties of the antisymmetric term $\vec{P}^{(a)}(\vec{r},t)$ are determined by the chromophore alone.

We start from Eq. (4.5)

$$\epsilon^{(a)}_{nlm}(\omega)=\int_{-\infty}^{\infty} ds e^{i\omega s}\int d\vec{\mathrm{r}}\, \mathrm{r}_n f^{(a)}_{lm}(\vec{\mathrm{r}};s), \tag{4.27}$$

and now arrive at an identity, cf. Eq. (4.25)

$$\begin{aligned} \int d\vec{\mathrm{r}}\, \mathrm{r}_n f^{(a)}_{lm}(\vec{\mathrm{r}};s) &= \int d\vec{\mathrm{r}}(-\mathrm{r}_n) f^{(a)}(-\vec{\mathrm{r}};s) = -\int d\vec{\mathrm{r}}\, \mathrm{r}_n[-f^{(a)}\vec{\mathrm{r}};s)] \\ &= \int d\vec{\mathrm{r}}\, \mathrm{r}_n f^{(a)}(\vec{\mathrm{r}};s). \end{aligned} \tag{4.28}$$

That is, the reference configuration of molecules without inversion symmetry sustains a non-zero vector

$$V_n \neq 0. \tag{4.29}$$

Because this non-zero value is consistent with the ground-state configuration, it does not include excited electronic states. That is, the linear term $\chi^{(1)}=-i\vec{V}\cdot\vec{\mathrm{k}}$ does not include dissipation. We conclude that for molecules without inversion symmetry, a polarization linear in the wave vector requires a different physical interpretation.

To this end, we note that our starting point is a Taylor-series expansion of the exponential factor $\exp(-i\vec{\mathsf{k}} \cdot \vec{r}) = 1 - i\vec{\mathsf{k}} \cdot \vec{r} + \ldots$ That is, the formalism requires that the electric fields (as well as polarization and electric displacements) are *complex*-valued. For an electromagnetic wave traveling along the z-axis, the electric field (as well as the magnetic field) are confined in the (x, y)-plane. We discussed in Sect. 1.1 that a multiplication with the imaginary unit i describes a rotation in the imaginary plane $z = re^{i\phi} \longrightarrow re^{i(\phi+\pi/2)}$. We conclude that the term $\vec{V} \cdot \vec{\mathsf{k}}$ is the 'mechanism' for a rotation of the electric field $\mathcal{E} = E_x + iE_i$:

$$\mathcal{E} \longrightarrow i\epsilon_0 \mathsf{k} \cdot l\mathcal{E}, \tag{4.30}$$

where we introduce the length scale $l = |\vec{V}|$. A rotation can be written in terms of vector product in terms of the Levi-Civita tensor: We thus arrive at the dielectric susceptibility

$$\epsilon_{lm} = \epsilon^{(0)}\delta_{lm} + l\sum_{n} \epsilon_{\mathrm{LC},lmn}\mathsf{k}_n. \tag{4.31}$$

This is equivalent with the polarization in terms of the vector product of the wave-vector and the electric field in Eq. (4.1).

Chiral molecules exist in a pair of configurations that are mirror images of each other (so-called enantiomers). A familiar example is alanine; we see that two versions of the molecule correspond to the left (latin: laevus—L) and right (latin: dextro—D). When we look at the back of our hands, the thumb of the left (right) hand is to the right (left) side. When we turn the right hand by 180 degrees, the thumb of the right hand is to the right; however, we look on the palm side of the right hand. Symmetric molecules are optically inactive. In large molecules, only a small pocket of atoms (ions) do not have inversion symmetry: this small pocket is referred to as *chromophore.*

Geometry considerations: Diatomic molecules have $2 \times 3 = 6$ degrees of freedom, which we identify with three degrees of freedom associated with the center of mass, two rotational degrees of freedom characterized two angles, and a vibrational degree of freedom. The latter depends on the distance between the two atoms (ions). For triatomic molecules, we have $3 \times 3 = 9$ degrees of freedom, we identify with three degrees of freedom associated with the center of mass, three rotational degrees of freedom characterized three (Euler-) angles, and three vibrational degrees of freedom (i..e, between atoms #A and #B , atoms #1 and #3, and atoms #2 and

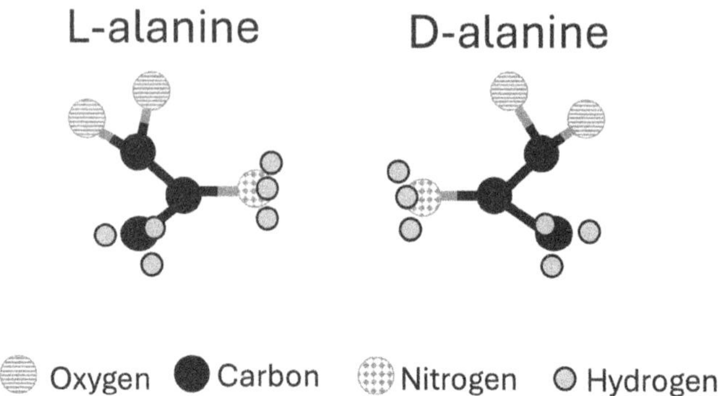

FIGURE 4.1
The chemical structure of alanine $CH_3CH(NH_2) - COOH$.

#3 described by distances $\xi_1 = |\vec{r}_A - \vec{r}_B|$, $\xi_2 = |\vec{r}_A - \vec{r}_C|$, and $\xi_3 = |\vec{r}_B - \vec{r}_C|$, respectively). The coupled vibrations can be described by normal coordinates, for example, the vibrational modes of the water molecule (H_2O), discussed in Sect. 2.4. Put simply, a plane is defined by the location of the three atoms. We can choose a coordinate system with two coordinates ξ_1 and ξ_2 in the plane (A,B,C) and the coordinate ξ_3 perpendicular to the plane. Since $\xi_3 = 0$ for all three atoms, the 'original' configuration of atoms coincides with its 'mirror image.' It follows that diatomic and triatomic molecules do not exhibit spatial dispersion.

The situation is different for molecules with *more than* three atoms (ions).[5] We discuss the case of a molecule with four atoms (ions). The distances between atoms do not uniquely define the molecular configuration: for fixed distances, there are two separate configurations obtained by coordinate inversion $\vec{r} \longrightarrow -\vec{r}$. For atoms labeled A, B, C, and D, we choose a coordinate system centered at the center of mass and introduce coordinates $\xi_1 = |\vec{r}_A|, \xi_2 = |\vec{r}_B|, \xi_3 = |\vec{r}_C| \geq 0$, and the positions of the three atoms A, B, and C define a plane (A,B,C). The position of particle D, $\vec{r}_D$ defines coordinates $-\infty < \xi_4, \xi_5, \xi_6 < +\infty$: the coordinates

[5]We follow the two papers by F. Hund: Symmetriecharactere von Termen by Systemen mit gleichen Partikeln in der Quantenmechanik, *Ann. Phys.* (Leipzig) **43**, 788 (1927), and Zur Deutung der Molekelspektren III—Bemerkungen über das Schwingungs- unde Rotationspektrum bei Molekeln mit mehr als zwei Kernen, *Ann. Phys.* (Leipzig) **43**, 805 (1927)—in German.

ξ_4 and ξ_5 are the coordinates of the projection of the position $\vec{r}_D$ on the plane (A,B,C), and ξ_6 is the 'height' of the position $\vec{r}_D$ above (below) the plane (A,B,C). We have the potential energy $U(\{\xi_i\}_{i=1,2,...,6})$. We fix the values $\xi_i = \bar{\xi}_i$ for $i = 1, 2, .., 5$ and find the potential energy as a function of ξ_6:

$$U'(\xi_6) = U(\{\xi_i = \bar{\xi}_i\}_{i=1,2,...,5}|\xi_6). \tag{4.32}$$

The potential energy is the identical when particle D is above or below the plane (A,B,C) so that $U'(\xi_6)$ is symmetric:

$$U'(-\xi_6) = U(\xi_6). \tag{4.33}$$

In general, the function $U'(\xi_6)$ has two minima

$$\xi_6 = \pm\xi_6^0 \quad \text{with} \quad \left.\frac{dU'}{d\xi_6}\right|_{\xi_6^0} = 0; \tag{4.34}$$

we ignore the special case when particle D lies in the plane (A,B,C) and $\xi_6^0 = 0$. The two minima can be identified with the right- and left-handed molecule. In a quantum mechanical description, the classical motion along the ξ_6 corresponds to states $|\psi_+\rangle$ and $|\psi_-\rangle$. The two states are energy eigenstates with well-defined values $H_0\,|\psi_\pm\rangle = E_0\,|\psi_\pm\rangle$; we say that the two states are *degenerate.*

4.2 Time-Dependent Fields

We start from the expression of the polarization

$$\vec{P} = \epsilon_0 l \left[-i\vec{\mathsf{k}} \times \vec{E}\right]. \tag{4.35}$$

We derived this expression by assuming an electromagnetic wave with wave vector $\vec{\mathsf{k}}$ (and frequency $\omega = c|\vec{\mathsf{k}}|$). This is equivalent with considering a Fourier component. We have seen in Sect. 1.1 that the derivative $f'(x) = df/dx$ corresponds to $-i\mathsf{k}\hat{f}(\mathsf{k})$. Thus, we have

$$-i\vec{\mathsf{k}} \times \vec{E} \quad\longleftrightarrow\quad \nabla \times \vec{E}. \tag{4.36}$$

so that the polarization is produced by the curl of the electric field

$$\vec{P} = \vec{P}_0 + \epsilon_0 l \nabla \times \vec{E}, \tag{4.37}$$

where we include a polarization $\vec{P}_0$ produced by a curl-free electric field $\vec{E}_0$ ($\nabla \times \vec{E}_0 = 0$).

Maxwell's equations relate electric to magnetic phenomena. In particular, the curl of an electric field, or *EMF*, is produced by a time-dependent magnetic field, $\nabla \times \vec{E} = -\partial \vec{B}/\partial t$ so that the polarization follows

$$\vec{P} = \vec{P}_0 - \epsilon_0 l \frac{\partial \vec{B}}{\partial t}. \tag{4.38}$$

If a time-dependent *magnetic* field can produce a polarization, the 'symmetry' between time-dependent electric and magnetic phenomena suggests that a time-dependent *electric* field can produce a magnetization[6]:

$$\vec{M} = \epsilon_0 l \frac{\partial \vec{E}}{\partial t}. \tag{4.39}$$

The RHS of Eqs. (4.38) and (4.39) are characterized by the same (or similar) length scale l; this suggests that the same (or similar) molecular mechanism present in optically active molecules produces both an electric and a magnetic dipole moment from time-dependent electromagnetic fields. That is, in optically active molecules, the induced dipole and magnetic moments are on equal footing.

We note that the electric field $\vec{E}$ and the polarization $\vec{P}$ are polar vectors, while the magnetic field $\vec{B}$ and the magnetization $\vec{M}$ are axial vectors. Thus, the vectors on the left- and right-hand sides of Eqs. (4.38) and (4.39) have a different 'character': a polar (axial) vector is generated by a time-dependent axial (polar) vector. This suggests that the mechanism of optical activity is 'outside' classical physics, where a polar (axial) vector is generally produced by a polar (axial) vector. For example, the time derivative of the linear momentum is equal to the force (both polar vectors) and the time-derivative of the angular momentum is generated by a torque (both axial vectors).

While Born did propose a classical theory of optical activity,[7] the theory is 'unwieldy,' or 'mühsam' as Rosenfeld pointed out, and a quantum-mechanical theory is required. This conclusion does *not*

[6] It is worthwhile to check the units. For Eq. () we have the LHS $[P] = \mathrm{C/m^2}$ and, using the unit for magnetic field $[B] = \mathrm{T} = \mathrm{N\,s/(C\,m)}$, the RHS $[\epsilon_0 l(\partial B/\partial t)] = \mathrm{C^2/(N\,m^2)} \cdot \mathrm{m} \cdot \mathrm{N/(C\,m)} = \mathrm{C/m^2}$; so that the LHS and RHS have the same unit. For Eq. (), we have the LHS $[M] = \mathrm{A/m}$ and the RHS $[\epsilon_0 l \partial E/\partial t] = \mathrm{C^2/(N\,m^2)} \cdot \mathrm{m} \cdot \mathrm{N/(C\,s)} = \mathrm{C/(m\,s)} = \mathrm{A/m}$, as it should.

[7] M. Born, Elektronentheorie des natürlichen optischen Drehvermögens isotroper and anisotroper Flüssigkeiten *Ann. Phys.* **55**, 177 (1918).

apply to frequency dispersion since in this case, the electric polarization (i.e., a polar vector) is generated by a time-dependent electric field (i.e., a polar vector). A quantum mechanical treatment of the vibrational motion of bonds only leads to insignificant modification since the expectation values of the coordinate and momentum operators $\langle x \rangle = \langle \psi|x|\psi \rangle$ and $\langle p \rangle$ $\langle \psi|p|\psi \rangle$ of a quantum-mechanical harmonic oscillator (with mass m and frequency ω_0) follows the laws of Newtonian mechanics, $\partial \langle x \rangle / \partial t = \langle p \rangle / m$ and $\partial \langle p \rangle / \partial t = -m\omega_0^2 \langle x \rangle$.

4.3 Quantum-mechanical Origin of Optical Activity

Evidently, electric polarization implies that molecules acquire a non-zero dipole moment. In classical physics, a dipole corresponds to charges $\pm q$ separated by the distance δr, $p = q\delta r$. The quantum-mechanical analog is the expectation value of the dipole operator,

$$q\delta\vec{r} \longrightarrow \langle \psi_0|\vec{p}|\psi_0 \rangle = \int d\vec{r}\, \psi_0^*(\vec{r})(q\vec{r})\psi_0(\vec{r}). \tag{4.40}$$

Here, the wave function $\psi_0(\vec{r})$ describes the ground state.

The quantum-mechanical nature of optical rotatory power implies interference of quantum states; that is, it involves both the ground state $|\, \psi_0\rangle$ as well as excited state $|\psi_\mu\rangle$, $\mu = 1, 2, 3, ...$ Since optical activity does not involve dissipation (damping), the system does *not* make a transition to an excited state. Rather the excited states are virtual states that affect the phase, but not the amplitude, of the wave function. The transitions from the ground state to a virtual states and from a virtual state to the ground state

$$|\psi_0\rangle \longrightarrow |\psi_\mu\rangle \;\; \text{(virtual)} \;\; \longrightarrow |\psi_0\rangle\,; \quad \mu = 1, 2, 3, ... \tag{4.41}$$

In quantum mechanics, transition probabilities are described by expectation values of operators; in the present context, the relevant operators correspond to the electric and magnetic dipole moments $\vec{p}$ and $\vec{m}$, that is,

$$\langle \psi_0|\vec{p}|\psi_0 \rangle \longrightarrow \sum_{\mu=1,2,...} c_\mu \langle \psi_0|\vec{p}|\psi_\mu \rangle \langle \psi_\mu|\vec{m}|\psi_0 \rangle\,. \tag{4.42}$$

The magnetic dipole moments (similar to the electric dipole moment) relate to the displacement of the atoms (ions) from their respective equilibrium positions. We conclude that the optical activity relates to the motion of atoms (ions). Thus, the starting point is an examination of the (possible) rotational and vibrational motion of atoms (ions) in molecules.

In the presence of a small perturbation (e.g., an external magnetic field) $\mathcal{H} = \mathcal{H}_0 + \delta\mathcal{V}$, the corresponding states are $\psi_+(\xi_6)$ and $\psi_-(\xi_6)$. The degeneracy is 'lifted' and the eigenstates are symmetric and antisymmetric superpositions

$$|\psi_s\rangle = \frac{1}{\sqrt{2}}\left[\,|\psi_+\rangle + |\psi_-\rangle\right], \tag{4.43}$$

$$|\psi_a\rangle = \frac{1}{\sqrt{2}}\left[\,|\psi_+\rangle - |\psi_-\rangle\right]. \tag{4.44}$$

so that $\mathcal{H}\,|\psi_{s,a}\rangle = E_{s,a}\,|\psi_{s,a}\rangle$ with energies $E_{s,a} = E_0 \pm \delta E/2$.[8] The time-dependent state is a linear position $|\psi\rangle = (1/\sqrt{2})[\,|\psi_+\rangle \exp(iE_+t/\hbar) + |\psi_-\rangle \exp(iE_-t/\hbar)]$. We assume $|\psi_+\rangle = |\psi_-\rangle = |\psi_0\rangle$ at time $t = 0$ and find

$$|\psi(t)\rangle = \frac{2}{\sqrt{2}}\,|\psi_0\rangle\, e^{iE_0t/\hbar}\cos\left(\frac{\delta E}{\hbar}t\right). \tag{4.45}$$

Since $\delta E < E$, $\cos([\delta E/\hbar]t)$ varies slowly compared to phase $\exp(i[E/\hbar]t)$. That is, the superposition of the symmetric and antisymmetric state produces a 'beat-like' behavior with a slowly varying amplitude

$$|\psi_0\rangle \quad \longrightarrow \quad |\psi_0(t)\rangle\cos\left(\frac{\delta E}{\hbar}t\right). \tag{4.46}$$

The quantum-mechanical description involves expectation values of operators $\mathcal{O}$ in the ground state

$$\langle\psi(t)|\mathcal{O}|\psi(t)\rangle \tag{4.47}$$

where the

$$|\psi(t)\rangle = e^{iEt/\hbar}\,|\psi\rangle\,, \qquad \langle\psi(t)| = \langle\psi|\,e^{-iEt/\hbar}. \tag{4.48}$$

[8] This is referred to as *degenerate* perturbation theory, see, for example, D. J. Griffiths, *Introduction to Quantum Mechanics* (Prentice Hall, Upper Saddle River, NJ, 1995).

If the operator is *symmetric* under coordinate inversion (such as the angular momentum operator) $\mathcal{O} = \mathcal{S}$, the expectation value is non-zero only for 'diagonal' terms $\langle\psi_s(t)|\mathcal{S}|\psi_s(t)\rangle$ and $\langle\psi_a(t)|\mathcal{S}|\psi_a(t)\rangle$. We find $\langle\psi_s(t)|\mathcal{S}|\psi_s(t)\rangle = \exp(i[E_s - E_s]t/\hbar))\,\langle\psi_s|\mathcal{S}|\psi_s\rangle = \langle\psi_s|\mathcal{S}|\psi_s\rangle$, that is, the expectation value is time-*independent*,

$$
\begin{aligned}
\langle\psi_s(t)|\mathcal{S}|\psi_s(t)\rangle &= \langle\psi_s|\mathcal{S}|\psi_s\rangle\,, \\
\langle\psi_a(t)|\mathcal{S}|\psi_a(t)\rangle &= \langle\psi_a|\mathcal{S}|\psi_a\rangle\,.
\end{aligned} \tag{4.49}
$$

If the operator is *antisymmetric* under coordinate inversion (such as the position or linear momentum operator) $\mathcal{O} = \mathcal{A}$, the expectation value is non-zero only for 'off diagonal' terms $\langle\psi_s(t)|\mathcal{A}|\psi_a(t)\rangle$ and $\langle\psi_a(t)|\mathcal{A}|\psi_s(t)\rangle$. We find $\langle\psi_s(t)|\mathcal{A}|\psi_a(t)\rangle = \exp(-i[E_s - E_a]t/\hbar))\,\langle\psi_s|\mathcal{A}|\psi_a\rangle = \exp(-i\delta Et/\hbar))\,\langle\psi_s|\mathcal{S}|\psi_a\rangle$, that is, the expectation value is time-*dependent*,

$$
\begin{aligned}
\langle\psi_s(t)|\mathcal{A}|\psi_a(t)\rangle &= \exp(-i\delta Et/\hbar))\,\langle\psi_s|\mathcal{A}|\psi_a\rangle\,, \\
\langle\psi_a(t)|\mathcal{A}|\psi_s(t)\rangle &= \exp(+i\delta Et/\hbar))\,\langle\psi_a|\mathcal{A}|\psi_s\rangle\,;
\end{aligned} \tag{4.50}
$$

this description applies to an oscillating electric dipole moment, which is the starting point for a quantum-mechanical description of frequency dispersion.

Optical rotatory power is explained by the interaction of light with matter; that is, it involves an oscillatory interaction $\mathcal{V}$ (different from the perturbation $\delta\mathcal{V}$ introduced above),

$$
H' = \mathcal{V}e^{i\omega t} + \mathcal{V}^\dagger e^{-i\omega t}, \tag{4.51}
$$

where $\mathcal{V}^\dagger$ is the adjoint operator.[9] This interaction changes the time evolution of the quantum state.[10] In addition, the interaction produces non-zero cross terms between ground and excited states In the context of a quantum-mechanical time evolution, we have the expression

$$
\langle\psi_\mu|\psi(t_0 + \Delta t)\rangle = \langle\psi_\mu|\psi(t)\rangle + \frac{\hbar}{i}\,\langle\psi_\mu|\mathcal{V}|\psi(t)\rangle \cdot \delta t. \tag{4.52}
$$

Since the quantum state is the eigenstate of the Hamiltonian in the absence of an electromagnetic field (i.e., $\mathcal{V} = 0$), the time-dependence is oscillatory (at least approximately) so that the theory of Fourier transforms, cf. Sect. 1.1. and the matrix

[9]The matrix elements are the complex conjugates $\mathcal{V}^\dagger_{\mu\mu'} = (\mathcal{V}_{\mu'\mu})^*$.

[10]It is instructive to recall the analogous expressions from elementary calculus. We have for a function at time $t_0+\Delta t$, $f(t_0+\Delta t) = f(t_0)+(df/dt)_{t=t_0}\cdot\Delta t$. The change for finite time differences $t > t_0$ is obtained by integration $f(t) = \int_{t_0}^{t}(df/ds)ds$.

element can be written as a division by the frequency difference $\langle\psi_\mu|\psi(t)\rangle = \hbar\,\langle\psi_\mu\mathcal{V}|\psi(t)\rangle\,/(E_\mu - E_0)$. As a result, the expectation value of an operator $\langle\psi(t)\mathcal{O}|\psi(t)\rangle$ must be modified. If both $\mathcal{O}$ and $\mathcal{V}$ are either symmetric or antisymmetric, the expectation is a sum of excited states μ of the general form.[11]

$$\frac{\langle\psi_s|\mathcal{V}|\psi_\mu\rangle\,\langle\psi_\mu|\mathcal{O}|\psi_s\rangle\,e^{i\omega t}}{(E_\mu - E_s) + \hbar\omega}. \tag{4.53}$$

If the operator $\mathcal{V}$ is symmetric (antisymmetric) and the operator is antisymmetric (symmetric), one finds E

$$\frac{\langle\psi_s|\mathcal{V}|\psi_\mu\rangle\,\langle\psi_\mu|\mathcal{O}|\psi_a\rangle\,e^{i[\delta E/\hbar+\omega]t}}{(E_\mu - E_s) + \hbar\omega}. \tag{4.54}$$

The expression Eq. (4.53) has the time-dependence $\exp(i\omega t)$, that is, the same dependence as the electromagnetic wave; in contrast, the expression Eq. (4.54) includes a phase shift $\Delta\phi = (\delta E/\hbar)\cdot t$. We have seen in Sect. 3.4 that the key mechanism for optical activity is the generation of a phase shift. We conclude that optical activity is explained by a molecular mechanism captured in Eq. (4.54) that connects two operators with opposite symmetry under inversion.

The rotation of the electric field is determined by the energy difference $\delta E = E_s - E_a$ between the energies of the symmetric and antisymmetric ground states. We expect that this difference is small so that the corresponding frequency $\delta E/h$ is small compared to the frequency of the incident electromagnetic wave, $\delta E/\hbar << \omega$. This is consistent with the estimate for the frequency difference between right- and left-circularly polarized light $\Delta\omega/\omega \simeq 10^{-6}$, cf. Eq. (3.50). If we assume light in the visible part of the spectrum, we find $\delta E/\hbar \simeq 10^{8}\,\mathrm{s}^{-1}$ so that $\delta E \simeq 10^{-26}\,\mathrm{J}$, or in terms of the energy scale $\mathcal{U} = 1\,\mathrm{kJ/mol}$, cf. Eq. (1.118) $\delta E/\mathcal{U} \simeq 10^{-6}$.

4.4 Annotated Bibliography

Spatial diffusion: Spatial diffusion is authoritatively discussed in L. D. Landau and E. M. Lifshitz, *Course in Theoretical Physics Vol 8—Electrodynamics of Continuous Media*

[11]See, for example, L. Rosenfeld. Quantenmechanische Theory der natürlichen Aktivität von Flüssigkeiten und Gasen, *Ann. Phys.* **52**, 161 (1928).

(Pergamon Press, London, 1960); a model of coupled oscillators is presented in J. J. Hopfield and D. G. Thomas, Theoretical and Experimental Effects of Spatial Dispersion on the Optical Properties of Crystals, *Phys. Rev.* **132**, 563 (1963).

5

Optically Active Molecules

The insight from Chapter 4 shows that optical rotatory power depends on the interplay between two seemingly unrelated 'features': it requires chiral molecules and the dynamics of atoms (ions) must be described in the framework of quantum mechanics. Even if the atoms (ions) are described quantum-mechanically, molecules with inversion symmetry cannot rotate the polarization axis. Likewise, if the dynamics of atoms (ions) were to obey the laws of Newtonian (classical) mechanics, no molecule could rotate the polarization axis, even those without inversion symmetry.

We sketch the outline of quantum-mechanical treatment of optical activity in Sections 5.1 and 5.2. The central point is that the necessary electronic transitions require that the configuration of molecules exist in virtual states. This is a purely quantum-mechanical property and is only possible in chiral molecules.

Many, perhaps even most, phenomena with purely quantum mechanical origin describe novel mechanisms and do not simply modify a classical phenomena.[1] Their explanations are counter-intuitive, and one seeks simple models that describe the quantum-mechanical in a simple model. Optical activity is no exception. In Sect. 5.3, we discuss the model by Drude that, in fact, precedes the development of quantum mechanics; we outline the physical origin of the Drude model in terms of coupled oscillators.

[1] Tunneling through a potential barrier is such a phenomena familiar to many readers: in a classical description, the α particle (He nucleus) is confined to the nucleus of a heavy ion. Radioactive decay requires a quantum-mechanical explanation.

DOI: 10.1201/9781003560944-5

5.1 Quantum Mechanical Theory

The interaction of light with matter necessarily involves some transition: it can be an electronic transition or a transition of atoms (ions) associated with vibrations or rotations. The strength of the coupling is inversely proportional to the mass. Since electron mass is small compared to the mass of an atom (ion) ($m_e/\mathcal{M} \simeq 1/1,836$), optical rotatory power, similarly as light absorption is largely due to electronic transitions. Thus optically active molecules are also good absorbers of light; however, the reverse is not true: many strong absorbers of light do not at all rotate the polarization axis of the incident light.

The transitions relevant to optical activity correspond to the near UV part of the electromagnetic spectrum and thus involve 'jumps' of electrons from the ground to an excited state. The contribution to optical activity from a transition to the IR band is small as it would involve vibrational motion of atoms (ions) in molecules.

The existence of discrete electronic states in atoms and molecules can only be explained in quantum theory of electronic degrees of freedom (either in terms of a Schrödinger equation or in terms of 'old' quantum theory familiar from Bohr model of the hydrogen atom). In this sense, a quantum-mechanical explanation is the basis for both light absorption and optical rotatory power.

For molecules, the electric susceptibility is characterized by the oscillator strength associated with the displacement of ions away from their equilibrium positions. In a quantum-mechanical description, the oscillator strength is expressed in terms of the expectation value of the electric dipole moment; however, a description in terms of classical mechanics description is equally possible. In this sense, a quantum-mechanical explanation of light absorption is not required.[2]

This provides the context in which we have to interpret the statement that 'optical activity requires a quantum-mechanical solution.'[3] That is, the quantum-mechanical nature relates to the dynamics of atoms (ions) in optically active molecules, which, in turn, 'facilitate' electronic transitions that eventually produce the

---

[2]We note, however, that the quantum mechanical nature of vibrations, as well as rotations, are apparent in the IR spectra of molecules.

[3]See Hecht, *Optics* 5th Ed. (Pearson, Boston, 2017), Sect. 8.10.

induced electric field perpendicular to the electric field of the incident electromagnetic wave. Since a permanent jump of the electron from the ground to an excited state is associated with light absorption, we conclude that the 'transition associated with optical activity must have some 'unique' character. In particular, the electron remains in its ground state so that the transition cannot be permanent and can instead possess only transient character.

The insight from spatial dispersion shows that the motion must define a certain handedness (either right- or left-handed) and is not possible in molecules *with* inversion symmetry. Many large, optically-active molecules possess inversion symmetry excerption for a small pocket of atoms (ions). This small pocket is referred to as *chromophore*. In the following, we refer to the 'chromophore' when we use the term 'molecule.' This usage is quite standard in the literature.

Electric polarization is determined by the displacement of electrons; the discussion can be reduced to a single valence as cooperative effects of electronic degrees of freedom are irrelevant. The trajectories of electrons is governed by the motion of atoms (ions); that is, characterizing the underlying motion of atoms in molecules is central for understanding the mechanism of optical activity.

We seen in Sect. 4.3 that optical rotation requires the involvement of an excited virtual state that provides the necessary expectation values of a symmetric and an antisymmetric operator

$$\langle \psi_s | \mathcal{V} | \psi_\mu \rangle \, \langle \psi_\mu | \mathcal{O} | \psi_a \rangle \, . \tag{5.1}$$

The involvement of a virtual state is the signature of a quantum-mechanical origin. The role of the virtual state becomes transparent in a path-integral formulation, cf. Sect. 1.6, where the time evolution of a quantum system is described in terms of a superposition of phases associated with 'possible' path of the corresponding classical system. In terms of classical quantities, the symmetry properties of quantum operators correspond to polar and axial vectors as the particles move along paths.

The choice of the polar vector is evidently the displacement of atoms (ions); we conclude that the axial vector is related to particle motion. The 'obvious' choice of the axial vector is the angular momentum $\vec{L} = \vec{r} \times \vec{p}$. For an ion, the angular momentum relates to the magnetic moment $\vec{m} = \gamma \vec{L}$, where $\gamma = q/2M$ is the gyromagnetic ratio. For quantum systems, the time evolution of a charged particle is governed by the interaction o with the electromagnetic potentials $(\phi, \vec{A})$. In contrast, the corresponding classical particle 'sees' the fields $(\vec{E}, \vec{B})$. Our discussion suggests that optical activity is governed by the interaction of the magnetic moment with the magnetic field of the incident light, $U = -\vec{m} \cdot \vec{B}$. We now outline how this classical view follows from the quantum-mechanical description. Our focus is understanding the physical origin of the various terms; accordingly, we use simpliications that must be corrected in a formal discussion. We follow the original quantum-mechanical solution by Rosenfeld from 1928.[4]

In the Lorentz gauge, the vector potential produces both electric and magnetic fields $\vec{E} = -\partial \vec{A}/\partial t$ and $\vec{B} = \nabla \times \vec{A}$. The fields and potential are oscillatory for an electromagnetic wave so that the vector potential can be written in terms of the electric field,

$$\vec{A}(\vec{r}, t) = -(i\omega)^{-1} \vec{E}(\vec{r}, t) = i\omega^{-1} \vec{E}_0 e^{i(\vec{k} \cdot \vec{r} - \omega t)}. \tag{5.2}$$

We start by considering the Hamiltonian of a single ion in a molecule. The starting point is the momentum $\vec{p} - (q/c)\vec{A}$ so that kinetic energy term of the Hamiltonian $(2M)^{-1}[\vec{p} - (q/c)\vec{A}]^2$ yields the interaction[5]

$$\mathcal{V} = -\frac{q}{2Mc} \vec{p} \cdot \vec{A}. \tag{5.3}$$

We use $\omega = c\mathsf{k}$ and obtain

$$\mathcal{V} = -\frac{q}{M\omega} \vec{p} \cdot \vec{E}_0 e^{i(\vec{\mathsf{k}} \cdot \vec{r} - \omega t)}. \tag{5.4}$$

Here $\vec{r}$ is displacement of the ion away from its equilibrium position. The insight from spatial diffusion from Chapter 4 suggests that a *non-local* term (i.e., a coordinate-dependence) is necessary. The displacement is small compared to the wavelength so that $\vec{\mathsf{k}} \cdot \vec{r} < 1$ so that we can approximate the coordinate dependence of the phase $\exp(i[\vec{\mathsf{k}} \cdot \vec{r} - \omega t]) \simeq \exp(i\omega t)[1 - i\vec{\mathsf{k}} \cdot \vec{r}]$. We obtain

$$\mathcal{V}' \simeq -i \frac{q}{M\omega} \vec{p} \cdot \vec{E}_0 \, (\vec{\mathsf{k}} \cdot \vec{r}) e^{-i\omega t}, \tag{5.5}$$

where we dropped to the local term $\mathcal{V} = -(q/M\mathsf{k})\, \vec{p} \cdot \vec{E}_0 e^{-i\omega t}$. Here, the imaginary unit i is the signature of the origin as a phase

[4]See L. Rosenfeld, Quantenmechanische Theory der natürlichen Aktivität von Flüssigkeiten und Gasen, *Ann. Phys.* **52**, 161 (1928).

[5]In a fully quantum mechanical calculation we have to consider both the terms $\vec{p} \cdot \vec{A}$ and $\vec{A} \cdot \vec{p}$ since the two operators, in general, do not commute. We ignore this 'subtlety' here as we are only interested in the physical origin of the various terms.

factor. Since the electric field is perpendicular to the direction of propagation $\vec{E} \cdot \vec{\mathsf{k}}$, we have, cf. Eq. (1.98)

$$\begin{aligned} \vec{E}_0(\vec{\mathsf{k}} \cdot \vec{r}) &= \vec{E}_0(\vec{\mathsf{k}} \cdot \vec{r}) - \vec{r}(\vec{E}_0 \cdot \vec{\mathsf{k}}) \\ &= \vec{\mathsf{k}} \times (\vec{r} \times \vec{E}_0) \end{aligned}$$

so that

$$\mathcal{V}' \simeq -i\frac{q}{M\omega}\vec{p} \cdot \left[\hat{\mathsf{k}} \times (\vec{r} \times \vec{E}_0)\right] e^{-i\omega t}. \tag{5.6}$$

We use the vector identity $\vec{a} \cdot (\vec{b} \times \vec{c}) = -\vec{c} \cdot (\vec{b} \times \vec{a})$,

$$\mathcal{V}' \simeq +i\frac{q}{M\omega}(\vec{r} \times \vec{E}_0) \cdot (\vec{\mathsf{k}} \times \vec{p})e^{-i\omega t}. \tag{5.7}$$

We note that the dependence on the position $\vec{r}$ enters via the scalar product with the wave vector $\vec{\mathsf{k}} \cdot \vec{r}$, which allows us to replace the coordinate vector $\vec{r}$ by the wave vector $\vec{k}$ and *vice versa*,

$$\vec{\mathsf{k}} \leftrightarrow \frac{1}{l^2}\vec{r}, \tag{5.8}$$

where l is a length scale. We write accordingly $\vec{r} \times \vec{E}_0 \to l^2(\vec{\mathsf{k}} \times \vec{E}_0)$. Since the curl corresponds to the vector product with the wave vector, we have $\nabla \times \vec{E} \to i\vec{\mathsf{k}} \times \vec{E}$. Oscillatory time-dependence $(\exp(-i\omega t))$ yields Faraday's law $[\nabla \times \vec{E} = \partial \vec{B}/\partial t]$ in the form $i\vec{\mathsf{k}} \times \vec{E}_0 = -i\omega \vec{B}_0$. We obtain

$$(\vec{r} \times \vec{E}_0) \to -l^2 \omega \vec{B}_0. \tag{5.9}$$

The interaction follows

$$\mathcal{V}' \simeq -i\frac{q}{M}l^2 \vec{B}_0 \cdot (\vec{\mathsf{k}} \times \vec{p})e^{-i\omega t}. \tag{5.10}$$

We write $l^2(\vec{\mathsf{k}} \times \vec{p}) = \vec{r} \times \vec{p}$, which equals the angular momentum

$$l^2(\vec{\mathsf{k}} \times \vec{p}) = \vec{L}. \tag{5.11}$$

Finally, the ratio q/M in terms of the gyromagnetic ratio $\gamma = q/2M$ so that $(q/M)\vec{L} = \vec{m}/2$. Since $\vec{B}_0 e^{-i\omega t} = \vec{B}(t)$, we arrive at the interaction

$$\mathcal{V}' \simeq -\frac{i}{2}\vec{B}(t) \cdot \vec{m}\,. \tag{5.12}$$

This is the desired result: embedding the atom (ion) in a magnetic field, the molecule sustains a rotational motion around a center, thereby creating a magnetic moment that interacts with the magnetic field associated with the incident light.

Similarly, the dependence on the angular momentum is not surprising as it is the 'obvious' choice of the axial vector characterizing molecular motion. A non-zero value of $\vec{L}$ requires a handedness of a molecule without inversion symmetry and thus is consistent with the explanation of rotatory power in the framework of spatial diffusion, cf. Chapter 4. In our 'derivation' (or rather outline), we introduced the length scale l characterizing the rotational motion of atoms (ions) about a center (e.g., the radius in the case of circular motion). This is consistent with our qualitative reasoning in Chapter 4; we arrived at the expression for optical activity in terms of the dimensionless quantity $g/\epsilon_0 = l\mathsf{k}$, cf. Eq. (3.51). We found $l \simeq 10^{-13}\,\text{m}$, which is a reasonable estimate for the displacement of atoms (ions).

We note that the interaction energy $\mathcal{V}'$ includes the factor i ($i^2 = -1$—the imaginary unit), similar to the expression for the electric permeability derived from spatial diffusion. At first blush, the similarity goes even further since the imaginary unit enters the expression via the expansion of the exponential dependence

$$e^{i\vec{\mathsf{k}}\cdot\vec{r}} \simeq 1 - i\vec{\mathsf{k}}\cdot\vec{r} + ... \tag{5.13}$$

We thus may conclude that the complex-valued nature in Eqs. (4.52) and (5.13) has similar physical origin. This conclusion is flawed, however. While we use a complex-valued notation for the electric (as well as the magnetic field, of course), this is only a matter of convenience. Maxwell's equations describe classical fields so that the fields can be expressed with purely real electric and magnetic fields, that is, with no imaginary component. Quantum mechanics is different, as the wave function *must* be complex valued, and a description in terms of a purely real-valued wave function does not capture essential features of a quantum system.

Seemingly unrelated properties go hand-in-hand to produce optical rotatory power. It starts with a chiral molecule that sustains both a symmetric and antisymmetric ground state. The ions are coupled to the electromagnetic field of the incident light: as a result, the quantum-mechanical state (vector or ket) of the ion acquires a time-dependent phase. In terms of paths, this coupling allows ions to 'explore' regions in coordinate-momentum, or phase-space that includes rotations about the center and is therefore associated with an angular momentum that defines right- or left-handedness.

Since atoms (ions) 'explore possible paths' rather than 'follow actual paths,' the electrons likewise do not undergo actual jumps from the ground to an excited state, but rather only undergo virtual transitions. An induced dipole moment requires an actual displacement of electrons associated with a (permanent) transition to an excited electronic state. That is, electrons remain at their equilibrium position during virtual transition so that their trajectories mirror those of the atoms (ions) in the molecule. As a result, virtual electronic transitions change the phase of the incident electric field without changing its magnitude.

While optical activity requires a quantum-mechanical explanation, it is nevertheless desirable to have a qualitative understanding based on classical mechanics. In framework of classical mechanics, the valence electron follows a trajectory such that electric and magnetic dipole moments are produced at the same time together.

5.2 Spiral Model of Optical Activity

A classical version of the quantum-mechanical theory of optical activity is, in principle, straightforward: possible paths of electrons in a quantum-mechanical description are (simply) replaced by actual paths of electrons in a classical description. However, this view is rather 'confusing' since the electron remains in the ground state; that is, in a classical description, the valence electron would be inert. These contradictory views cannot be resolved in a description based on classical physics. This incompatibility reflects the quantum-mechanical nature of optical activity but is generally ignored in classical descriptions.

The classical motion must generate both an induced electric and a magnetic moment at the same time, which requires an 'intertwined' combination of both linear and circular motion of the valence electron. This description fits the motion along a spiral. Interestingly, this model was proposed by Drude *prior* to the development of quantum mechanics.[6] Since a valence electron follows the trajectory imposed by the motion of ions, the spiral reflects the

[6] P. Drude, Über magnetooptische Erscheinungen, *Göttinger Nachrichten* 366 (1892); see also P. Drude, *Theory of Optics* English translation (Longmans, New York, 1902), p. 400.

'shape' of molecules.[7] The length scale l associated with optical activity characterizes the radius of the spiral, the enclosed area of a loop πl^2.

If the generation of the spiral trajectory is key mechanism of optical activity, it is left to show that no additional mechanism is necessary to rotate the polarization of linearly polarized light. As pointed out by others, this is rather challenging task because the orientations of molecules, and thus the corresponding spirals, are random and appropriate averages must be calculated. We follow 'common practice' and do not explore details of the spatial averaging; it is sufficient to examine a particular geometry to get a 'feel' for the mechanism.

Space dependence: We follow the description in Feynman's Lectures.[8] We consider light linearly polarized along the x-axis and assume that the spiral is aligned (mostly) with the x-axis, cf. Fig. 5.1. The electric field causes the electron to move up and down the spiral. The spiral motion will cause the electron to also move *into* the page (along the $-y$-axis) at $z = -l$ and *out-of* the page (along the $-y$-axis) at $z = +l$. One might expect that the corresponding components of the electric field cancel out each other since the two components cancel out each other; i.e., they are out-of-phase and differ by the phase $\Delta\pi = \pi$. This is not the case however since the two electric fields have a phase difference which is slightly different $\Delta\phi = \pi + \mathsf{k}l$.

Time-dependence: We follow the description in the review by Kauzmann et al.[9] We assume that the spiral is tilted slightly away from the x-axis so that the normal vector $\hat{n}$ has components $n_x \simeq 1$, $n_y < 0$ and $n_z > 0$ with $|n_y|, |n_z| << 1$. We further assume incident light that is right-circularly polarized, and consider the situation when the electric field is directed along the $+x$-axis. We consider the situation when the electron is the position with coordinates $(x = -0, y = -l, z = 0)$, cf. Fig. 5.2. Since the electron is *negatively*

[7]The formulation in Feynman Vol. I provides a fairly typical example. The reader is told to '[c]onsider an asymmetric molecule in the shape of a spiral. [...] Molecules need not actually be shaped like a corkscrew in order to exhibit optical activity, but this simple shape which we shall take as a typical example of those that do not possess reflection symmetry.'

[8]R. P. Feynman, R. B. Leighton, and M. Sands, *Feynman Lectures on Physics* Vol. 1 (Addison-Wesley, Reading MA, 1965), Sect. 33.5.

[9]W. J. Kauzmann, J. E. Walter, and H. Eyring, Theories of Optical Rotatory Power, *Chem. Rev.* **26**, 339–407 (1940).

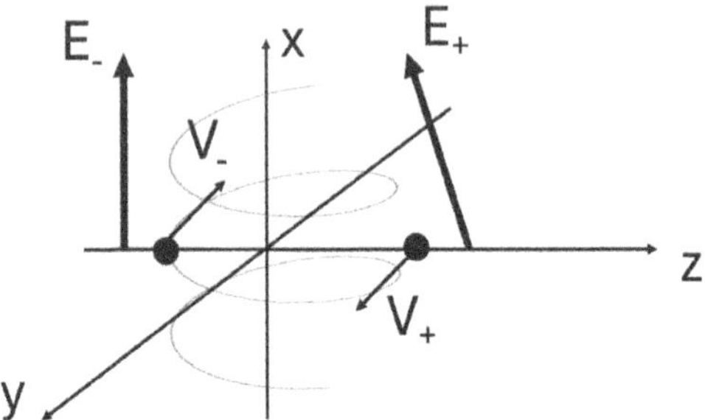

FIGURE 5.1
The electromagnetic wave travels along the $+z$-axis. The spiral is directed (mostly) along the x-axis. The electron along the $-y$-axis at $z = -l$ and along the $+y$-axis at $z = +l$. The electric fields at these two locations point in (slightly) different directions due to the time delay associated with the wave traveling across the spiral.

charged so that force $\vec{F} = (-e)\vec{E}$ is directed along the $-x$-axis, and the electron tends to move in this direction; the induced dipole moment is directed along the $+x$-axis, $\vec{p}_0 = p_0\hat{x}$. This response is consistent with general dielectric matter; optical activity describes the response to time-dependent fields.

For right-circular polarization, the time derivative of the magnetic field is directed in the same direction as the electric field, that is, along the $+x$-axis, so that the rate of change of the magnetic flux is positive $d\Phi_B/dt = \pi l^2(\partial B/\partial t) > 0$. A time-dependent magnetic flux produces an electromotive force (*EMF*) that moves the electron along a loop. Lenz' law shows the electron will move along the loop in *counterclockwise* direction such that the current associated with the flow produces a magnetic field directed along the $-x$-direction, cf. Fig. 5.1. That is, a time-dependent magnetic field displaces the electron along the $+x$-axis so that polarization is directed along the $-x$-axis, $\delta\vec{p} \sim -(\partial B/\partial t)\hat{x}$.

The time-dependent electric field is associated with a displacement current that produces a magnetic field $\vec{B}'$ directed in clockwise direction when viewed from the $+y$-axis. The magnetic force exerted on the moving electron is then directed toward the center of the spiral, and thus sustains the motion of the electron along the spiral, cf. Fig. 5.3. When viewed from the $+y$-axis, the spiral

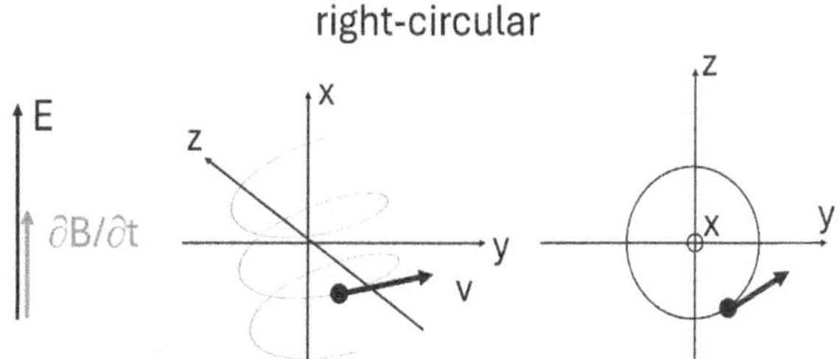

FIGURE 5.2
For right-circularly polarized light, the electric field $\vec{E}$ and time derivative $\partial\vec{B}/\partial t$ point in the $+x$-direction. The electron travels in *counterclockwise* direction to produce an induced magnetic field that points along the $-x$-axis following Lenz' law.

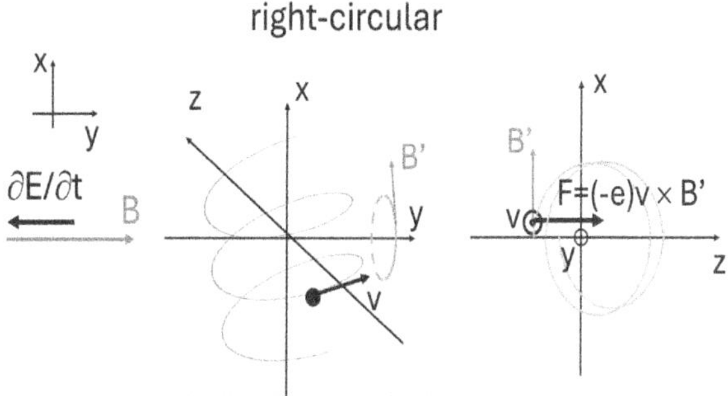

FIGURE 5.3
For right-circularly polarized light, the magnetic field $\vec{B}$ points in the $+y$-direction. and time derivative $\partial\vec{E}/\partial t$ points in the $-y$-direction. The displacement current associated with the time-varying electric generates the magnetic field $\vec{B}'$ that keeps the electron traveling in a (distorted) circular motion when viewed from the $+y$-axis.

resembles a distorted circle so that the current associated with the moving electron produces a magnetic moment along the $+y$-axis, $\delta\vec{m} \sim +(\partial E/\partial t)\hat{y}$, consistent with Eq. (4.38).

We now consider the case for left-circularly polarized light, cf. Fig. 5.4. The time-derivative of the magnetic field is directed in the opposite direction as the electric field, i.e., along the $-x$-axis, so that the rate of change of the magnetic flux is negative $d\Phi_B/dt =$

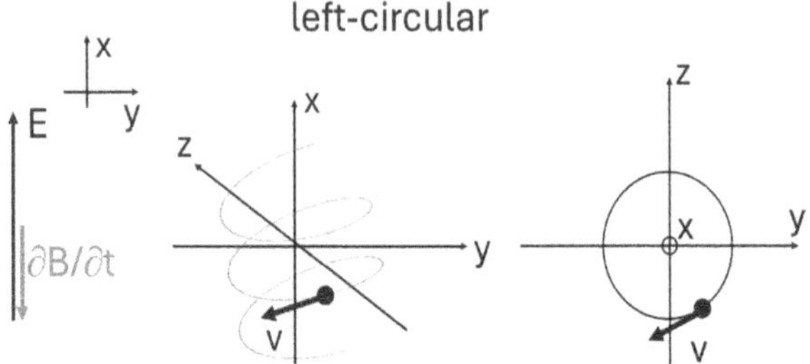

FIGURE 5.4
For left-circularly polarized light, the electric field $\vec{E}$ points in the $+x$-direction and time and the time derivative $\partial\vec{B}/\partial t$ points in the $-x$-direction. The electron travels in *counterclockwise* direction to produce an induced magnetic field that point along the $-x$-axis following Lenz' law.

$\pi l^2(\partial B/\partial t) < 0$. Lenz' law shows the electron will move along the loop in *clockwise* direction such that the current associated with the flow of produces a magnetic field directed along the $+x$-direction, cf. Fig. 5.3. That is, a time-dependent magnetic field displaces the electron along the $-x$-axis so that polarization is directed along the $-x$-axis, $\delta\vec{p} \sim +(\partial B/\partial t)\hat{x}$.

The time-dependent electric field is associated with a displacement current that produces a magnetic field $\vec{B}'$ directed in clockwise direction when viewed from the $-y$-axis. The magnetic force exerted on the moving electron is then directed toward the center of the spiral, and thus sustains the motion of the electron along the spiral, cf. Fig. 5.5. When viewed from the $+y$-axis, the spiral resembles a distorted circle so that the current associated with the moving electron produces a magnetic moment along the $-y$-axis, $\delta\vec{m} \sim -(\partial E/\partial t)\hat{y}$.

For time-varying electric fields associated with an incident light, the electron will move up and down the spiral. The rotational motion along loops will create a magnetic dipole moment along the spiral axis. It follows that the electric and the magnetic dipole moment have a non-zero overlap $\vec{p} \cdot \vec{m} \neq 0$. The overlap depends on the 'pitch' of the spiral, that is, direction of the normal vector of the spiral $\hat{n}$.

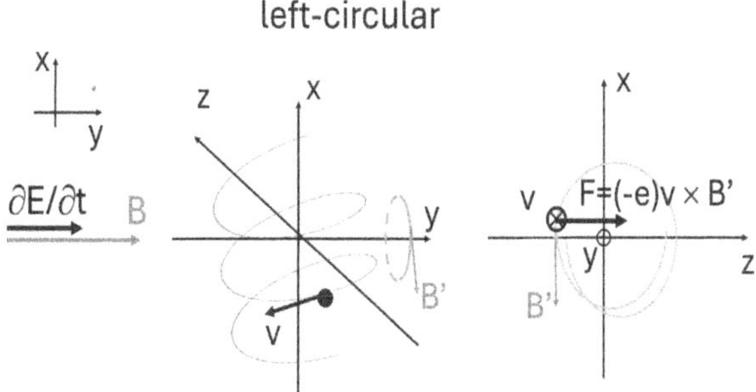

FIGURE 5.5
For left-circularly polarized light, the magnetic field $\vec{B}$ and time derivative $\partial\vec{E}/\partial t$ point in the $+y$-direction. The displacement current associated with the time-varying electric generates the magnetic field $\vec{B}'$ that keeps the electron traveling in a (distorted) circular motion when viewed from the $+y$-axis.

5.3 Coupled Oscillator Model of Optical Activity

The classical description in the preceding section leads us to discuss optical activity in terms of spiral motion. Indeed, this view can be found in numerous explanation. The treatment in Feynman Vol. I can (perhaps) serve as a fairly typical example. The reader is told to '[c]onsider an asymmetric molecule in the shape of a spiral. [...] Molecules need not actually be shaped like a corkscrew in order to exhibit optical activity, but this simple shape which we shall take as a typical example of those that do not possess reflection symmetry.'

It is quite obvious that most molecules do not have the structure of a spiral as is certainly the case for alanine. The spiral trajectory is generated by molecules, that is the internal dynamics of atoms (ions) of molecules. At first blush, this is a surprising conclusion since the dynamics is characterized by normal modes. The normal modes are independent of each other; the vibration along one mode does not affect the vibrations along another mode; see Sect. 1.5 for a discussion.

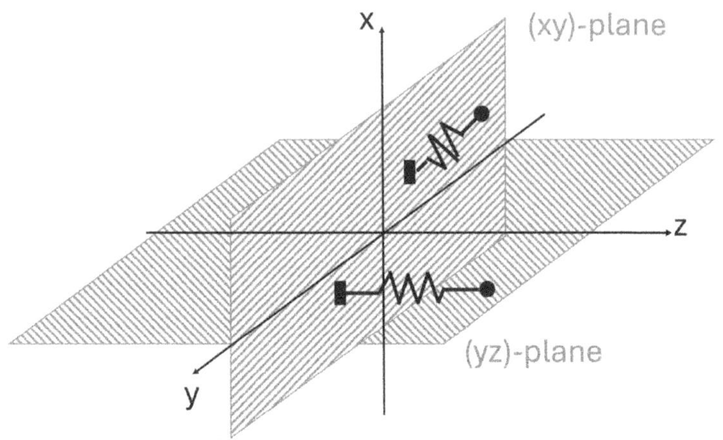

FIGURE 5.6
The two-coupled oscillator model: the oscillators are anisotropic and are restricted to oscillate in the $(xy)-$ and $(yz)-$ planes, respectively. Molecules without inversion sustain a coupling between the oscillators generating the spiral trajectory for the valence electron.

This discussion must be modified for molecules *without* inversion symmetry. As we have seen, in that case, a molecule can sustain a coupling between vibrations (oscillations) along two perpendicular axes (say along x- and y-axes). We say that the oscillators are *anisotropic*. Depending on the nature of the forcing, the trajectory defines either a clockwise or counterclockwise direction (details are not important for our discussion).

The motion along a spiral is obtained if the two oscillators are not only along two orthogonal direction in a single plane, but in addition, are displaced in the direction along the normal to the plane. For a spiral with the long axis directed along the x-coordinate, we arrive at a model with two coupled oscillators vibrating in the (xy)- and (yz)-plane, respectively that are displaced along the x-axis, cf. Fig. 5.6. This model was originally proposed by Kuhn.

The displacements of atoms (ions) govern the response of molecules to an incident electromagnetic wave. It explains the (frequency) dispersion of dielectric matter, cf. Sect. 2.5; in particular, the electric susceptibility is determined by the change of the bond length $\langle x \rangle \sim (\omega_\mu^2 - \omega^2 - i\omega\gamma_\mu)^{-1}$ and the oscillator strength f_μ, cf. Eq. (2.29). Optical rotatory power is similarly governed by the displacements of the two oscillators and a coupling strength:

$$A_{\mu\mu'}\left(\frac{1}{[\omega_\mu^2-\omega^2]}-\frac{1}{[(\omega'_\mu)^2-\omega^2]}\right), \tag{5.14}$$

for frequencies of the incident several orders of magnitude higher than vibrational frequencies $\omega >> \omega_\mu, \omega'_\mu$. Since the electron remains in the ground state, damping can be ignored $\gamma_\mu = 0$.

5.4 Annotated Bibliography

Optical activity: We find older papers and reviews of optical activity (in chronological order) M. Born, Über die ntürlichen optischen Aktivitäten von Fl"ussigkeiten und Gasen, *Z. Phys.* **XVI**, 251 (1915), M. Born, Electronentheorie des natürlichen optischen Derhevermögens isotroper and anistroper Flüssigkeiten, *Ann. Phys.* (Leipzig, Germany) **55**, 177 (1918), E. U. Condon, Theories of Optical Rotatory Power, *Rev. Mod. Phys.* **9**, 432 (1937), W. J. Kauzmann, J. E. Walter, and H. Eyring, Theory of Optical Rotatory Power, *Chem Rev.* **26**, 339 (1940), D. D. Fitts and J. G. Kirkwood, The Optical Rotatory Power of Helical Molecules, *Proc. Natl.* 251 (1915), *Acad. Sci. (USA)* **42**, 33 (1956), S. Chandrashekar, Simple Model for Optical Activity, *Am. J. Phys.* **24**, 503 (1956), L. L. Jones and H. Eyring, A Model for Optical Rotation, *J. Chem. Ed.* **38**, 601 (1961), J. A. Schellman, Circular Dichroism and Optical Rotation, *Chem Rev.* **75**, 323 (1975).

Coupled-oscillator model: The coupled oscillator model was developed mainly by Werner Kuhn in a series of papers in the 1920s and 1930s. W. Kuhn, Quantitive Verhältnisse und Beziehungen bei der natürlichen optischen Aktivität, *Z. Phys. Chem.* **B4**, 14 (1929), W. Kuhn, The Physical Significance of Optical Rotatory Power, *Trans. Faraday Soc.* **216**, 293 (1930).

6

Optical Activity in a Nutshell

As pointed out in the Introduction, the straightforward description of optical rotatory power is in stark contrast to the challenge posed to explain its underlying physical mechanism. The challenge stems, partially at least, from the mathematical 'language' used in physics. This is, of course, not limited to the topic of this book; it is difficult to grasp the *beauty* of Maxwell's equations without a firm knowledge of vector calculus (curl, divergence, and all that).

As the late Nobel Prize winner Eugene Wigner wrote '[t]he enormous usefulness of mathematics in the natural sciences is something bordering on the mysterious and there is no rational explanation for it. It is not at all natural that the "laws of nature" exists, much less than that man [or woman, UZ] is able to discover them. The miracle of the appropriateness of the language of mathematics for the formulation of the laws of physics is a wonderful gift which we neither understand nor deserve.'[1] Perhaps, in some ('metaphysical') sense, vector calculus 'exists' for us to understand electromagnetism.

In this vein, complex numbers 'exist' for us to understand quantum mechanics. In fact, quantum mechanics cannot be formulated in terms of real numbers: the wave function is complex-valued, that is, it has a real and an imaginary part, and the time-dependent Schrödinger equation is equivalent with a diffusion equation in imaginary time.[2]

Optical rotatory power refers to properties of electric fields, which is a vector quantity. The addition of two vectors follows

[1]E. Wigner, *The Unreasonable Effectiveness of Mathematics in the Natural Sciences*, in Communications in Pure and Applied Mathematics, **13**, No. 1 (February 1960).

[2]There are attempts of description of quantum mechanics in terms of real numbers; see, for example, D. Garisto, Physicists Take the Imaginary Numbers of Quantum Mechanics, *Quanta Magazine*, downloaded 11/11/2025. The author agrees with statement by Valtko Vedral (Oxford University) cited in the article: 'You can write them down whichever way you like, but it's unavoidable that they have to multiply exactly as though they were complex numbers.'

DOI: 10.1201/9781003560944-6

vector rules (we add separately the x- and y-components). We are in the fortuitous situation that these vector rules are identical to the rules of adding two complex numbers (we add separately the real and imaginary parts). That is, complex numbers (and thus the use of the imaginary unit $i = \sqrt{-1}$) provide the mathematical framework to describe the quantum-mechanical mechanism that underlies the rotation of the electric field.

Optical activity relates to properties of the molecule in its ground state. A molecule with a single, well-defined ground state is 'inert' and cannot sustain any time-dependence in the framework of both classical and quantum mechanics. Such a molecule does not possess rotatory power; this applies to most molecules. Molecules without inversion symmetry have two ground states that correspond to the right- and left-handed configuration. The degeneracy is 'lifted' by a small interaction resulting in a symmetric and an antisymmetric state. In the framework of quantum mechanics, this splitting of energy levels allows tor *virtual* transitions even in the ground state, analogous to beats of nearly-degenerate vibrating strings. Virtual transitions correspond to a change in the phase of the quantum-mechanical state that is governed by the vector potential (the scalar potential can be set equal to zero). In turn, the quantum phase determines the phase of the complex-valued electric (as well as magnetic) fields. In terms of real (rather than complex-valued) fields, a change in the phase corresponds to a rotation of the fields around the direction of propagation.

While optical activity is a quantum effect at a fundamental level, some aspects can be described in a classical model of coupled anisotropic oscillators. The existence of two orthogonal oscillators is common and describes the vibrational motion of atoms (ions) in molecules both with and without inversion symmetry. The unique feature of molecules without inversion symmetry is the coupling between oscillators. A classical particle 'sees' the electric and magnetic fields: optical activity is explained by the response of an electron traveling on a spiral to time-dependent electric and magnetic fields.

Optical activity is a 'small' effect; this reflects the fact the atom remains in the ground state. In particular, the valence electron does not undergo a transition to an excited state. Due to much bigger mass, the trajectories of atoms (ions) is a small fraction of the radius of an electron orbit in an excited state.

Index

For Product Safety Concerns and Information please contact our EU representative GPSR@taylorandfrancis.com
Taylor & Francis Verlag GmbH, Kaufingerstraße 24, 80331 München, Germany

www.ingramcontent.com/pod-product-compliance
Ingram Content Group UK Ltd.
Pitfield, Milton Keynes, MK11 3LW, UK
UKHW020013110726
473185UK00003B/53

* 9 7 8 1 0 3 2 9 1 0 1 6 1 *